CAMBRIDGE LIBRARY COLLECTION

Books of enduring scholarly value

Darwin

Two hundred years after his birth and 150 years after the publication of 'On the Origin of Species', Charles Darwin and his theories are still the focus of worldwide attention. This series offers not only works by Darwin, but also the writings of his mentors in Cambridge and elsewhere, and a survey of the impassioned scientific, philosophical and theological debates sparked by his 'dangerous idea'.

History of Quadrupeds

Thomas Pennant (1726-98) was a keen geologist, naturalist and antiquary, who wrote a number of successful travel books about the British Isles as well as works on science. Linnaeus supported his election to the Royal Swedish Society of Sciences in 1757, and in 1767 he became a Fellow of the Royal Society. His work in zoology also earned him an honorary degree. His History of Quadrupeds (1793), aimed to promote natural history among a wider readership, originated in an informal index to John Ray's Synopsis of 1693. In his preface, Pennant acknowledges the monumental Histoire naturelle by the Comte de Buffon, as well as works by Klein (1751), Brisson (1756), and particularly the work of Linnaeus, though Pennant strongly disagreed with Linnaueus's classification of primates as including humans with apes. Pennant's two-volume book, beautifully illustrated with over 100 engravings, provides an accessible overview of the state of zoological classification at the end of the eighteenth century. Charles Darwin owned a copy and had it sent to him in South America during the Beagle voyage.

Cambridge University Press has long been a pioneer in the reissuing of out-of-print titles from its own backlist, producing digital reprints of books that are still sought after by scholars and students but could not be reprinted economically using traditional technology. The Cambridge Library Collection extends this activity to a wider range of books which are still of importance to researchers and professionals, either for the source material they contain, or as landmarks in the history of their academic discipline.

Drawing from the world-renowned collections in the Cambridge University Library, and guided by the advice of experts in each subject area, Cambridge University Press is using state-of-the-art scanning machines in its own Printing House to capture the content of each book selected for inclusion. The files are processed to give a consistently clear, crisp image, and the books finished to the high quality standard for which the Press is recognised around the world. The latest print-on-demand technology ensures that the books will remain available indefinitely, and that orders for single or multiple copies can quickly be supplied.

The Cambridge Library Collection will bring back to life books of enduring scholarly value (including out-of-copyright works originally issued by other publishers) across a wide range of disciplines in the humanities and social sciences and in science and technology.

History of Quadrupeds

Volume 2

Thomas Pennant

CAMBRIDGE
UNIVERSITY PRESS

CAMBRIDGE UNIVERSITY PRESS

Cambridge, New York, Melbourne, Madrid, Cape Town, Singapore,
São Paolo, Delhi, Dubai, Tokyo

Published in the United States of America by Cambridge University Press, New York

www.cambridge.org
Information on this title: www.cambridge.org/9781108005173

© in this compilation Cambridge University Press 2009

This edition first published 1793
This digitally printed version 2009

ISBN 978-1-108-00517-3 Paperback

HISTORY

of

QUADRUPEDS,

The Third Edition.

Vol. II.

P. Mazell Sculp.

LONDON:

Printed for B. & J. WHITE, Fleet Street.

MDCCXCIII.

HISTORY

OF

QUADRUPEDS.

Six cutting teeth, and two canine, in each jaw.
Five toes before; five behind.
In walking refts on the hind feet, as far as the heel.

Urfus. *Plinii lib.* viii. *c.* 36.
Α ρχτ Θ ·. *Oppian Cyneg.* iii. 139.
Urfus. *Gefner quad.* 941. *Agricola, An.
Subter.* 486. *Raii fyn. quad.* 171.
Niedzwiedz. *Rzaczinfki Polon.* 225.
Bâr. *Klein quad.* 82. *Schwenckfelt The-
riotroph.* 131. *Ridinger Wild. Thiere.*
31. *Arct. Zool.* i. No. 20.

Urfus niger, cauda concolore. *Briffon
quad.* 187.
Urfus, cauda abrupta. *Lin. fyft.* 69.
Biorn. *Faun. fuec. No.* 19.
L'Ours. *De Buffon,* viii. 248. *tab.* xxxi.
xxxii. *Schreber,* cxxxix. cxl. Lev.
Mus.

208. Brown.

B. with a long head: fmall eyes: fhort ears, rounded at the
top: ftrong, thick, and clumfy limbs: very fhort tail:
large feet: body covered with very long and fhaggy hair, various
in its color: the largeft of a rufty brown: fome from the confines of
Ruffia, black, mixed with white hairs, called by the *Germans, filver
bar* ; and fome (but rarely) are found in *Tartary* of a pure white.

Inhabits the north parts of *Europe* and *Afia*, the *Alps* of *Switzer-land*, and *Dauphiné*; *Arabia* *, *Japan* †, and *Ceylon* ‡; and the northern parts of *North America*; and extends to the *Andes* of *Peru*: Doctor *Shaw* informs us, it is alfo found in *Barbary*. They muft have been very plentiful, for *Pliny* fays that *Domitius Ænobarbus* produced at one of the fhews a hundred *Numidian* bears, and as many *Æthiopian* hunters §. The brown bears are fometimes carnivorous, and will deftroy cattle, and eat carrion; but their general food is roots, fruits, and vegetables: will rob the fields of peafe; and when they are ripe, pluck great quantities up; beat the peafe out of the hufks on fome hard place, eat them, and carry off the ftraw: they will alfo, during winter, break into the farmer's yard, and make great havock among his ftock of oats: are particularly fond of honey.

They live on berries, fruits, and pulfe, of all kinds; and feed much on the black mulberry: are remarkably fond of potatoes, which they very readily dig up with their great paws: make great havock in the fields of *maiz*; and are great lovers of milk and honey. They feed much on herrings, which they catch in the feafon when thofe fifh come in fhoals up the creeks; which gives their flefh a difagreeable tafte; and the fame effect is obferved when they eat the bitter berries of the *Tupelo*.

Bears ftrike with their fore foot like a cat; feldom or never ufe their mouths in fighting; but feizing the affailant with their paws, and preffing him againft their breaft, almoft inftantly fqueeze him to death.

The females, after conception, retire into the moft fecret

---

* *Forfkal*, iv.   † *Kæmpfer, Hift. Japan*, i. 126.   ‡ *Knox, Hift. Ceylon*. 20.
§ Lib. viii. c. 36.

places;

places; leaft, when they bring forth, the males fhould devour the young: it is affirmed for fact, that out of the feveral hundred bears that are killed in *America*, during winter, (which is their breeding feafon) that fcarcely a female is found among * them; fo impenetrable is their retreat during their pregnancy: they bring two, rarely three, young at a time: the cubs are deformed, but not a fhapelefs mafs, to be licked into fhape, as the antients pretended†. The cubs even of the brown bears are of a jetty blacknefs, and often have round their necks a circle of white. The flefh of a bear in autumn, when they are moft exceffively fat, by feeding on acorns, and other maft, is moft delicate food; and that of the cubs ftill finer; but the paws of the old bears are reckoned the moft exquifite morfel: the fat white, and very fweet: the oil excellent for ftrains, and old pains.

The latter end of autumn, after they have fattened themfelves to the greateft degree, the bears withdraw to their dens, where they continue for a great number of days in total inactivity, and abftinence from food, having no other nourifhment than what they get by fucking their feet, where the fat lodges in great abundance. In *Lapland* they pafs the long night in dens lined warmly with a vaft bed of mofs, in which they roll themfelves, fecure from the cold of the fevere feafon ‡. Their retreats are either in cliffs of rocks; in the deepeft receffes of the thickeft woods; or in the hollows of antient trees, which they afcend and defcend with furprizing agility: as they lay in no winter provifions, they

* Out of 500 bears that were killed in one winter, in two counties of *Virginia*, only two females were found, and thofe not pregnant. *Lawfon*, 117.

† *Hi funt candida informifque caro, paulo muribus major, fine oculis, fine pilo; ungues tantum prominent: hanc lambendo paulatim figurant.* Plinii lib. viii. c. 36.

‡ *Fl. Lap.* 313. The mofs is a variety of the *Polytrichum Commune.*

are

are in a certain fpace of time forced from their retreats by hunger; and come out extremely lean : multitudes are killed annually in *America*, for the fake of their flefh, or fkins, which laft
makes a confiderable article of commerce.

B. with a long pointed nofe, and narrow forehead: the cheeks
and throat of a yellowifh brown color : hair over the whole
body and limbs of a gloffy black, fmoother and fhorter than that
of the *European* kind.

They are ufually fmaller than thofe of the old world : yet Mr.
Bartram gives an inftance of an old he-bear killed in *Florida*, which
was feven feet long, and, as he gueffed, weighed four hundred
pounds.

Thefe animals are found in all parts of *North America*, from
Hudfon's Bay to the fouthern extremity ; but in *Louifiana* and the
fouthern parts they appear only in the winter, migrating from
the north in fearch of food. They fpread acrofs the northern
part of the *American* continent to the *Afiatic* ifles. They are
found in the *Kurilfki* iflands, which intervene between *Kamt-*
fchatka and *Japan*, *Jefo*, *Mafima*, which lies north of *Japan*, and
probably *Japan* itfelf ; for *Kæmpfer* fays, that a few fmall bears are
found in the northern provinces.

It is very certain that this fpecies of bear feeds on vegetables.
Du *Pratz*, who is a faithful as well as intelligent writer, relates,
that in one fevere winter, when thefe animals were forced in multitudes from the woods, where there was abundance of animal food,
they rejected that, notwithftanding they were ready to perifh with
hunger, and migrating into the lower *Louifiana*, would often break

Polar Bear ___ N.º 210.

into the courts of houfes. They never touched the butchers meat which lay in their way, but fed voracioufly on the corn or roots they met with.

White bear. *Martin's Spitſberg.* 100. *App.* xxvi. *Arƈt. Zool.* i. No. 18. *Egede Greenl.* 59. *Ellis voy.* 41. *Crantz Greenl.* i. 73. *Barentz voy.* 18. 45. *La Hontan voy.* i. 235. *Cateſby Carolina,* *Urſus albus. Martenſii. Klein quad.* 82. L'Ours blanc. *Briſſon quad.* 188. *De Buffon,* xv. 128. *Schreber,* cxli. Lev. Mus.

210. Polar.

B. with long head and neck: ſhort round ears: end of the nofe black: vaft teeth; hair long, foft, white, tinged in fome parts with yellow: limbs of great fize and ftrength: grows to a vaft fize: the ſkins of fome are thirteen feet long.

This animal is confined to the coldeft part of the globe: it has been found as far as navigators have penetrated northwards, above *lat.* 80. The frigid climates only feem adapted to its nature. It is unknown, except on the ſhores of *Hudſon's Bay, Greenland,* and *Spitzbergen.* The north of *Norway,* and the country of *Meſen,* in the north of *Ruſſia,* are deſtitute of them: but they are met with again in great abundance in *Nova Zembla,* and from the river *Ob,* along the *Siberian* coaſt, to the mouths of the *Jeneſei,* and *Lena,* but are never feen far inland, unleſs they lofe their way in mifts; none are found in *Kamtſchatka,* or its iſlands.

They have been feen as far fouth as *Newfoundland;* but they are not natives of that country, being only brought there accidentally on the iſlands of ice.

During fummer the white bears are either refident on iſlands of ice, or paſſing from one to another: they fwim admirably, and

PLACE.

MANNERS.

can.

can continue that exercife* fix or feven leagues; and dive with
great agility. They bring two young at a time: the affection
between the parents and them is fo ftrong, that they would die rather than defert one another. Their winter retreats are under
the fnow†, in which they form deep dens, fupported by pillars
of the fame, or elfe under fome great eminence beneath the fixed
ice of the frozen fea.

They feed on fifh, feals, and the carcafes of whales; and on
human bodies, which they will greedily difinter: they feem very
fond of human blood; and are fo fearlefs as to attack companies
of armed men, and even to board fmall veffels. When on land,
they live on birds, and their eggs; and, allured by the fcent of
the feals flefh, often break into and plunder the houfes of the
Greenlanders: their greateft enemy in the brute creation is the
Morfe ‡, with whom they have terrible conflicts, but are ge-
rally worfted; the vaft teeth of the former giving it a fupe-
riority.

The flefh is white, and faid to tafte like mutton: the fat is
melted for train oil, and that of the feet ufed in medicine; but
the liver is very unwholefome, as three of *Barentz*'s failors expe-
rienced, who fell dangeroufly ill on eating fome of it boiled.

One of this fpecies was brought over to *England*, a few years
ago: it was very furious, almoft always in motion, roared loud,
and feemed very uneafy, except when cooled by having pail-fulls
of water poured on it.

Callixenus Rhodius §, in his defcription of the pompous pro-

* *La Hontan,* i.   † *Egede,* 60.   ‡ *Egede, Greenl.* 60. 83.
§ As quoted by *Athenæus, lib.* v. *p.* 201.

ceffion

ceffion of *Ptolemæus Philadelphus* at *Alexandria*, fpeaks of *one great white Bear*, Αρκτος λευκη μεγαλη μια, among other wild beafts that graced the fhew: notwithftanding the local fituation of this fpecies at prefent, it is poffible that *Ptolemy* might procure one; whether men could penetrate, in thofe early times, as far as the prefent refidence of thefe *Arctic* animals, I will not venture to affirm, nor to deny; but fince my friend, the Hon. *Daines Barrington* *, has clearly proved the intenfe cold that in former ages raged in countries now more than temperate, it is moft probable that in thofe times they were ftocked with animals natural to a rigorous climate; which, fince the alteration, have neceffarily become extinct in thofe parts: the *Polar* bear might have been one; but that it was the fpecies meant by *Callixenus* is clear to me, by the epithet μεγαλη, or *great*, which is very applicable to it; for the white *Tartarian* land bear (which *Ptolemy* might very eafily procure) differs not in fize from the black or brown kind, but the bulk of the other is quite characteriftic.

PIED LAND BEARS.

Land bears, fometimes fpotted with white; at other times wholly white; are fometimes obferved on the parts of *Ruffia* bordering on *Siberia*, in a wandering ftate, fuppofed to have ftrayed out of the lofty fnowy mountains, which divide the two coun- tries †.

* *Phil. Tranf. vol.* lviii. *p.* 58.   † Doctor *Pallas.*

Quickhatch.

211. WOLVERENE.

Quickhatch *Catefby Carolina, App.* xxx.
Carcajou, or Quickhatch. *Dobbs Hud-
fon's Bay*, 40.
Quickhatch, or Wolverene. *Ellis Hud-
fon's Bay*, 42.   *Clerk's voy.* ii. 3.
Edw. 103.
Urfus lufcus. U. cauda elongata, cor-
pore ferrugineo, roftro fufco, fronte

plagaque laterali corporis. *Lin. fyft.*
71. *Arct. Zool.* 1. Nº 21.
Urfus. *Freti Hudfonis.* U. caftanei coloris, cauda unicolore, roftro pedibufque fufcis. *Briffon quad.* 188.
Schreber, cxliv.
Le Glouton. *De Buffon, Supplem.* iii.
244. LEV. MUS.

B. with a black fharp-pointed vifage: fhort rounded ears, almoft hid in the hair: hairs on the head, back, and belly,
reddifh, with black tips, fo that thofe parts appear, on firft
fight, quite black: fides of a yellowifh brown, which paffes in
form of a band quite over the hind part of the back, above the
tail: on the throat a white fpot: on the breaft a white mark, in
form of a crefcent: legs very ftrong, thick and fhort, of a deep
black: five toes on each foot*, not deeply divided: on the fore
foot of that I examined were fome white fpots: the bottom of
the feet covered very thickly with hair: refts, like the bear, on its
foot, as far as the firft joint of the leg; and walks with its back
greatly arched: claws ftrong and fharp, white at their ends: tail
cloathed with long coarfe hairs; thofe at the bafe reddifh, at the
end black: fome of the hairs are fix inches long: length from nofe

* Mr. *Edwards* obferved only four toes on the fore feet of the animal he defcribes. My defcription is taken from an entire fkin, in very fine prefervation,
communicated to me by the late Mr. *Afhton Blackburne,* of *Orford, Lancafhire,*
who, with indefatigable induftry and great judgment, enriched the cabinets of
his friends with the rareft natural productions of that continent: as this work
has profited fo greatly by that gentleman's labors, it would be ungrateful to
omit my acknowlegements.

Wolverene __ N.° 211.

to tail twenty-eight inches: length of the trunk of the tail feven inches, but the hairs reach fix beyond its end: the tail in Mr. *Edwards*'s figure not quite accurate: it is corrected in that which is borrowed from his admirable work. The whole body is covered with very long and thick hair, which varies in color, according to the feafon.

Inhabits *Hudfon's-Bay*, and *Canada*, as far as the ftraights of *Michilimakinac*.

A moft voracious animal: flow of foot, fo is obliged to take its prey by furprize: in *America* is called the *Beaver-Eater*, for it watches thofe animals as they come out of their houfes, and fometimes breaks into their habitations, and devours them.

In a wild ftate is vaftly fierce; a terror to both wolf and bear, which will not prey on it when they find it dead*, perhaps on account of its being fo very foetid, fmelling like a pole-cat: makes a ftrong refiftance when attacked; will tear the ftock from the gun, and pull the traps it is caught in to pieces: burrows†, and has its den under ground. Mr *Graham*, long refident in *Hudfon's Bay*, has affured me, that it will lurk on a tree, and drop on the deer which pafs beneath, and faften on them till the animals are quite exhaufted.

Charlevoix, in *Hift. Nouv. France*, v. 189, gives the name of this animal *(Carcajou)* to our 189th fpecies, the *Puma*, or Brown Panther of *N. America*.

---

* *Clerk California*, ii. 3.　　† *La Hontan's voy.* i. 62.

In conformity to the opinion of that refpectable naturalift Doctor *Pallas*, I unite the *Woolverene* and *Glutton*. I do not alter my defcription of the latter ; but add both that and the fynonyms; fubmitting to future times the propriety or impropriety of uniting thefe animals: there being diftinctions that even now leave me very undetermined.

212. GLUTTON.

Gulo. *Olaii Magni gent. Septentr.* 138.
Gulo, vielfrafs. *Gefner quad.* 554. *Klein quad.* 83. *tab.* v.
Rofomak. *Rzaczinfki Polon.* 218. *Bell's Travels*, i. 235.
Muller's Rufs Samlung. iii. 549, 550. *Ritchkoff Topogr. Orenb.* i. 295.
Jerf, Fieldfrofs. *Strom Sondmor.* 152. *Pontop. Norway*, ii. 22. *Scheffer's Lapland,* 134.

Hyæna. *Briffon quad.* 169. *Yfbrandts Ides Trav. Harris's Coll.* ii. 923.
Muftela gulo. M. pedibus fiffis, corpore rufo-fufco, medio dorfi nigro. *Lin. fyft.* 67. *Zimmerman.* 311.
Jarf, Filfrefs. *Faun. fuec.* No. 14.
Jæerven. *Gunner's Act. Nidros.* iii. 143. *tab.* iii.
Le Glutton. *De Buffon*, xiii. 278.

B. with a round head : thick blunt nofe : fhort ears, rounded, except at the tip: limbs large: back ftrait; marked the whole length with a tawny line: tail fhort and very full of hair: the hair in all other parts black, finely damafked or watered like a filk, and very gloffy; but fometimes varies into a browner color. *Klein* attributes to it five toes on each foot: that which Mr. *Zimmerman* defcribes, had but four, very thickly covered with hair.

Size. The length of one which was brought from *Siberia*, and kept alive at *Drefden*, was a yard and eight inches : the height from the top of the head was nineteen inches. Mr. *Zimmerman* de-

fcribes

fcribes another, rather leffer than the former, which was fhot near *Helmftedt*, in *Wolfenbuttle*. Its length was three feet three: its height before fifteen inches; behind, fixteen: the tail fix inches.

Inhabits *Lapland*, the northern and eaftern parts of *Siberia*, and *Kamtfchatka*. Thofe of *Kamtfchatka* differ, and vary to white and yellowifh, and their fkins are efteemed by the natives before the black: they fay, that the heavenly beings wear no other garments. The women wear the paws of the white fort in their hair: and efteem a fkin as the moft valuable prefent which their hufbands or lovers can make.

They are exceffively voracious; that which was confined at *Drefden* would eat thirteen pounds of flefh in a day, and not be fatisfied. The report of their filling themfelves fo full, as to be obliged to go between two trees to force out part of the food, feems to be fabulous.

Like the Lynx, it lurks on the boughs of trees, and will fall on any animal which paffes by, faften on, and deftroy it. Its game is chiefly deer; and about the *Lena*, horfes. Is capable of being made tame.

It differs from the bear by its lean habit; by not lying in-active in winter; and by its living entirely on animal food. It is alfo more bold, voracious, and cunning.

The *Ruffians* call it *Rofomak*; the *Kamtfchatkans*, *Timmi*; and the *Koratfki*, *Haeppi*. An animal, called by the *Greenlanders*, *Amanki*, is faid to be found in their country, which is fuppofed to be the Glutton; but as *Greenland* is deftitute of wood, I fup-pofe their *Amanki*, or *Amarok*, to be a fabulous animal*.

---

* See *Crantz Hift. Greenland.*

C 2

Raccoon

Raccoon. *Lawfon Carolina*, 121. *Catef-by Carolina, App.* xxix.
Mapach, feu animal cuncta prætentante manibus. *Hernandez, Nov. Hifp.* 1. *Nieremberg.* 175.
Vulpi affinis Americana. *Raii fyn. quad.* 179. *Sloane Jamaica,* ii. 329.
Coati. *Worm. Muf.* 319.
Coati. Urfus cauda annulatim varie-grata. *Briffon quad.* 189.
Urfus Lotor. U. cauda annulata, fafcia per oculos tranfverfali nigra. *Lin. fyft.* 70. *Arct. Zool.* i. N° 22.
Le Raton. *De Buffon,* viii. 337. *tab.* xliii. *Schreber,* cxliii.
Raccoon. *Kalm's Travels. Forfter's Tr.* i. 96. 208. *tab.* 11. LEV. MUS.

B. with a fharp-pointed black nofe: upper jaw the longer: ears fhort, and rounded: eyes furrounded with two broad patches of black: from the forehead to the nofe a dufky line: face, cheeks, and chin, white: upper part of the body covered with hair, afh-colored at the root, whitifh in the middle, and tipt with black: tail very bufhy, annulated with black: toes black, and quite divided. Sometimes this animal varies: I have feen one entirely of cream color*.

Inhabits the warm and temperate parts of *America:* found alfo in the mountains of *Jamaica,* and in the ifles of *Maria,* between the S. point of *California* and *Cape Corientes,* in the S. Sea†: an animal eafily made tame; very good-natured and fportive, but as unlucky as a monkey; almoft always in motion; very inquifitive, examining every thing with its paws; makes ufe of them as hands: fits up to eat: is extremely fond of fweet things, and ftrong liquors, and will get exceffively drunk: has all the cunning of a fox: very deftructive to poultry; but will eat all forts

* LEV. MUS.　　† *Dampier's voy.* i. 276.

of

of fruits, green corn, &c.: at low water feeds much on oyfters; will watch their opening, and with its paw fnatch out the fifh; fometimes is caught in the fhell, and kept there till drowned by the coming in of the tide: fond alfo of crabs: climbs very nimbly up trees: hunted for its fkin; the fur next to that of the beaver, being excellent for making hats.

Wha Tapoua Row. *White's Bot. Bay,* 278.

214. NEW HOLLAND.

B. of the fame external form as the *American* Raccoon except the ears, which are pointed: fix cutting teeth in the upper jaw; two ? in the lower: back of a dark grey; growing lighter on the fides: belly of a fine brown: tail as long as the body, covered with long hair; the lower part near the end is naked, and has a pre-henfile quality like fome fpecies of monkies, or the common *Opof-fum.*

Inhabits *New Holland.*

PLACE.

Six

XXI. BADGER.   Six cutting teeth, two canine, in each jaw.

Five toes before, five behind: very long ftrait claws on the fore feet.

A tranfverfe orifice between the tail and the anus.

215. COMMON.   Meles. *Plinii lib.* viii. *c.* 38. *Gefner quad.* 327.
Meles, five Taxus. *Raii fyn. quad.* 185.
Meles, Taxus, Taffus, Blerellus; Jazwiec, Borfuk. *Rzaczinfki Polon.* 233.
Coati cauda brevi, Coati grifeus, Taxus, Meles, Tax. *Klein quad.* 73.
Dachs. *Kramer Auftr.* 313.
Meles pilis ex fordidé albo et nigro variegatis veftita, capite tæniis alternatim albis et nigris variegato. *Briffon quad.* 183.
Le Blaireau, ou Taifon. *De Buffon,* viii. 104. *tab.* vii.
Urfus meies. U. cauda concolore, corpore fupra cinereo, fubtus nigro, faf-cia longitudinali per oculos aurefque nigra. *Lin. fyft.* 70.
Meles unguibus anticis longiffimis. Graf-fuin. *Faun. fuec.* No. 20. *Br. Zool.* i. 64. *Br. Zool. illuːr. tab.* lii. *Schreber,* cxlii. LEV. MUS.

B. with fmall eyes: fhort rounded ears: fhort thick neck: with nofe, chin, lower fides of the cheeks, and middle of the forehead, white: ears and eyes inclofed in a pyramidal bed of black: hairs on the body long and rude; their bottoms a yellowifh white, middle black, ends afh-colored: throat, breaft, belly, and legs black: tail covered with long hairs, colored like thofe on the body: legs very fhort and thick: claws on the fore feet very long: a fœtid white matter exudes from the orifice beneath the tail: animal of a very clumfy make.

SIZE.   The length is commonly two feet fix inches from the nofe to the origin of the tail; of the tail fix inches: the weight from fifteen to thirty-four pounds. The laft is rare; but I met with, in the winter of 1779, a male of that weight.

Inhabits

Inhabits moſt parts of *Europe,* as far north as *Norway**, and *Ruſſia*; and the *ſtep* or deſert beyond *Orenburgh,* in the *Ruſſian Aſiatic* dominions; in *Great Tartary,* and in *Siberia* about the river *Tom,* and even about the *Lena,* but none in the north; inhabits alſo *China,* and is often found in the butchers ſhops in *Pekin,* the *Chineſe* being fond of them†. A ſcarce animal in moſt countries: ſeldom appears in the day; confines itſelf much to its hole: is indolent and ſleepy: generally very fat: feeds by night; eats roots, fruits, graſs, inſects, and frogs: not carnivorous: its fleſh makes good bacon: runs ſlowly; when overtaken comes to bay, and defends itſelf vigorouſly: its bite hard and dangerous: burrows under ground, makes ſeveral apartments, but forms only one entrance from the ſurface: hunted during night, for the ſkin, which ſerves for piſtol furniture; the hair, for making bruſhes to ſoften the ſhades in painting. The diviſion of this ſpecies into two, *viz.* the ſwine and the dog badger, unneceſſary, there being only one.

Arct. Zool. i. No. 23.

β. AMERICAN.

B. with a white line from the tip of the noſe, paſſing between the ears, to the beginning of the back, bounded on each ſide, as far as the hind part of the head, with black, then by a white one, and immediately between that and the ears is another of black: hair long: back colored like that of the common badger: ſides yellowiſh: belly cinereous: thighs duſky: tail covered with long dirty yellow hairs, tipped with white; the end duſky.

* *Pontop. hiſt. Norway,* ii. 28. † *Bell's travels,* ii. 83.

The

The legs were wanting in the ſkin I took my deſcription from. M. *de Buffon*'s deſcription, taken from a ſtuffed animal * brought from *Terra di Librador*, will ſupply that defect: he ſays there were only four toes on the fore feet; but he ſuſpects (as I imagine was the caſe) that the fifth might have been rubbed off in ſtuffing.

Deſcribed from a ſkin from *Hudſon's-Bay*, found in a furrier's ſhop in *London:* it was leſs than that of the *European* badger: the furrier ſaid, he never met with one before from that country. *Kalm*† ſays, he ſaw the *European* badger in the province of *Penſylvania*, where it is called the *Ground Hog* ‡; and this proves to be no other, varying very little from it.

216. INDIAN.

B. with a ſmall head, and pointed noſe: ſcarcely any external ears; only a ſmall prominent rim round the orifice, which was oval: color of the noſe and face, a little beyond the eyes, black: crown, upper part of the neck, the back, and upper part of the tail, white, inclining to grey: legs, thighs, breaſt, belly, ſides, and under part of the tail black.

Five toes on each foot; the inner ſmall: claws very long and ſtrait.

SIZE:

Length from noſe to tail about two feet: tail four inches: hair ſhort and ſmooth.

* He calls it *Le Carcajou*. Suppl. iii. 242. *tab*. xlix.

† *Kalm's travels, Forſter's tranſl*. i. 189.

‡ M. *Briſſon* deſcribes a white Badger, with a yellowiſh white belly, and alſo much inferior in ſize to that of *Europe*, which M. *Reaumur* received from *New York*. Vide *Briſſon quad*. 185.

Inhabits

Inhabits *India:* feeds on flesh: is playful, lively, and good- PLACE.
natured: sleeps rolled up, with its head between its hind legs;
sleeps little in the day: refused all commerce with the *English*
badger which was turned to it, and lived some time in the same
place: climbs very readily over a division in its cage.

XXII. OPOS-
SUM.

Two canine teeth in each jaw.

Cutting teeth unequal in number in each jaw *.

Five toes on each foot: hind feet formed like a hand, with a diftinct thumb.

Tail very long, flender, and ufually naked.

217. VIRGINIAN.

Tlaquatzin. *Hernandez Mex.* 330. *Nieremberg,* p. 136. and fig. 136.
Tajibi. *Marcgrave Brafil.* 222. *Raii fyn. quad.* 182. 185.
Semi-vulpa. *Gefner quad.* 870. *Icon. An.* 90.
Opoffum. *Ph. Tr. abridg.* ii. 884. *tab.* xiii; iii. 593; and v. 169. 177. *Lawfon Carolina,* 120. *Beverley's Virginia,*
135. *Catefby Carolina, App.* xxix. *Rochefort Ant.les,* i. 283.
Fara, ou Ravall? *Gumilla, Orenoque,* iii. 258. *Arct. Zool.* i. No. 24.
Le Manicou. *Feuilleè obf. Peru.* iii. 206.
Wood-rat. *Du Pratz Louifiana,* ii. 65.
Didelphis marfupialis. D mammis octo intra abdomen? *Lin fyft.* 71. *Amœn. Acad.?* i. 561. LEV. MUS.

O. with long fharp-pointed nofe: large, round, naked, and very thin ears, black, edged with pure white: fmall, black, lively eyes: long ftiff hairs each fide the nofe, and behind the eyes: face covered with fhort foft white hairs: fpace round the eyes dufky: neck very fhort; its fides of a dirty yellow: hind part of the neck and the back covered with hair above two inches long; foft, but uneven; the bottoms of a yellowifh white, middle part black, ends whitifh: fides covered with dirty and dufky hairs; belly, with foft, woolly, dirty white hair: legs and thighs black: feet dufky: claws white: bafe of the tail clothed with long hairs, like thofe on the back; reft of the tail covered

* This fpecies has eight cutting teeth in each jaw. *Tyfon.*

with

Virginian Opossum — N° 217

with fmall fcales; the half next the body black, the reft white: it has a difagreeable appearance, looking like the body of a fnake, and has the fame prehenfile quality as that of fome monkies: body round, and very thick: legs fhort: on the lower part of the belly of the female is a large pouch, in which the teats are lodged, and where the young fhelter as foon as they are born.

The ufual length of the animal is, from the tip of the nofe to the bafe of the tail, about twenty inches; of the tail twelve inches.

Inhabits *Virginia, Louifiana, Mexico, Brafil,* and *Peru*: is very deftructive to poultry, and fucks the blood without eating the flefh: feeds alfo on roots and wild fruits: is very active in climbing trees: will hang fufpended from the branches by its tail, and, by fwinging its body, fling itfelf among the boughs of the neighbouring trees: continues frequently hanging with its head downwards: hunts eagerly after birds and their nefts: walks very flow: when purfued and overtaken, will feign itfelf dead: not eafily killed, being as tenacious of life as a cat: when the female is about to bring forth, fhe makes a thick neft of dry grafs in fome clofe bufh at the foot of a tree, and brings four, five, or fix young at a time.

As foon as the young are brought forth, they take fhelter in the pouch, or falfe belly, and faften fo clofely to the teats, as not to be feparated without difficulty: they are blind, naked, and very fmall when new-born, and refemble *foetufes*: it is therefore neceffary that they fhould continue there till they attain a perfect fhape, ftrength, fight, and hair; and are prepared to undergo what may be called a fecond birth: after which, they run into this pouch as into an afylum, in time of danger; and the parent carries them about with her. During the time of this fecond

SIZE.

PLACE.

MANNERS.

FALSE BELLY.

geftation,

geftation, the female fhews an exceffive attachment to her young, and will fuffer any torture rather than permit this receptacle to be opened; for fhe has power of opening or clofing it by the affiftance of fome very ftrong mufcles.

The flefh of the old animals is very good, like that of a fucking pig: the hair is dyed by the *Indian* women, and wove into garters and girdles: the fkin is very fœtid.

M. *de Buffon* feems not to be acquainted with this animal, but has compiled an account of its manners, and collected the fynonyms of it. The figures * which he has given belong to the following fpecies, as does the defcription.

---

218. Molucca.

Carigue, ou Saragoy. *De Laet*, 485.
Carigueya. *Marcgrave*, 223.
Mus Marfupialis, *Beutel ratze. Klein quad.* 59.
Vulpes major putoria cauda tereti & glabra? *Barrere France Æquin.* 166.

Philander orientalis fœmina. *Seb. Muf.* i. 61. *tab.* xxxvi. *fig.* 1. 2. xxxviii. *fig.* 1.
Sarigue, ou l'Opoffum. *De Buffon*, 311. x. *tab.* lxv. lxvi. *Schreber*, cxlvi. A. B. Lev. Mus.

O. with long, oval, and naked ears: mouth very wide: over each eye is an oblong fpot of white: lower fide of the upper jaw, throat, and belly, of a whitifh afh-color: reft of the hair of a cinereous brown, tipt with tawny, darkeft on the back: tail long as the body; near the bafe covered with hair, the reft naked: claws hooked.

On the belly of the female is a pouch, in which the young (like thofe of the former) fhelter. *Marcgrave* found fix young within the pouch of the *Carigueya*, which I confider as the fame animal. It had ten cutting teeth above, and eight below.

* The figure in the firft edition was very indifferent, I have therefore changed it for the very faithful one in the *Phil. Tranf.*

3

Length

Length from nose to tail, ten inches. The tail exceeds the length of head and body. Its whole figure is of a much more slender and elegant make than the former.

The tail pulverised, and taken in a glass of water, is reckoned in *New Spain* a sovereign remedy against the gravel, colic, and several other disorders.

This genus is not confined to *America*, as M. *de Buffon* supposes; who combats the opinion of other naturalists on this subject with much warmth: but the authority of *Piso*, *Valentyn*, and of *Le Bruyn*, who have seen it both in *Java* and in the *Molucca Isles*, and of numbers of collectors in *Holland*, who receive it frequently from those places. This and N° 219 are proofs of what I advance. It is also met with in *New Holland.*

This species is found in great numbers in *Aroe* and *Solor*: It is called in the *Indies*, *Pelandor Aroe*, or the *Aroe Rabbet*. They are reckoned very delicate eating; and are very common at the tables of the Great, who rear the young in the same places in which they keep their rabbets. It inhabits also *Surinam,* and the hot parts of *America.*

Seba figures and describes, in his 1st vol. 64. *tab.* xxxix. an Opossum under the name of *Philander maximus orientalis fœmina.* It has a pouch like the former: is much larger: seems to have a longer and more slender tail: has broader ears; has a dusky spot over each eye, and is of a darker color. It feeds on fruits: was brought from *Amboina*, where it is called *Coes Coes**.

---

* *In Indiis orientalibus, idque solum, quantum hactenus constat, in* Amboina, *similis Bestia* (Carigueya) *frequens ad felis magnitudinem accedens, mactata ab incolis comeditur, si rite preparetur, nam alias fœtet. Nomen illi* Cous Cous *inditum.* Piso India, 323.

I am

I am unacquainted with this species, so leave these two conjoined till I receive fuller information.

Much is wanted to complete the natural history of this genus.

219. JAVAN.   Filander. *Le Bruyn voy. East Indies*, ii. 101. *tab.* ccxiii. *Ed. Angl.*

O. (according to *Le Bruyn*'s figure) with a narrow fox-like head: upright pointed ears: a brown stripe passing through the eyes: fore legs very short: five toes on the fore feet; three only on the hind, two of which are very strong; the outmost slender and weak; and found on dissection to consist internally of two bones, closely united, with two weak claws bursting out of the skin *: tail thick, shorter than the body.

In the upper jaw are six cutting teeth; in the lower two, which are formed like those of squirrels: no canine teeth †.

On the belly is a complete pouch, like the *Virginian* kind: hair on the body rude: face seemingly that of a hare.

PLACE.   Discovered first by Mr. *Le Bruyn*, who saw in *Java* several in an inclosure along with rabbets: they burrowed like them; leaped in their pace; preserved their young in the pouch, which would often peep out when the old ones were still.

The fidelity of *Le Bruyn*'s figure has been since confirmed by the specimens sent from *Java* into *Holland.*

* *Pallas* in act. acad. *Petrop. pars* ii. 229. *tab.* ix*.   † The same.

Mus fylveftris Americanus *Scalopes* dictus. *Seb. Muf.* i. 46. *tab.* xxxi. *fig.*
1. 2.
Philander faturate fpadiceus in dorfo,
in ventre dilutè flavus, pedibus albi-
cantibus. *Briffon quad.* 211.
Didelphis murina. D. cauda femipilofa,
mammis fenis. *Lin. fyft.* 72.
La Marmofe. *De Buffon,* x. 336. *tab.*
lii. liii. *Schreber,* cxlix.

O. with long broad ears, rounded at the end, thin and naked: eyes encompaffed with black : face, head, and upper part of the body, of a tawny color: the belly yellowifh white: the feet covered with fhort whitifh hair : toes formed like thofe of the *Virginian:* tail flender, covered with minute fcales, from the tip to within two inches of the bafe, which are cloathed with hair. Length from nofe to tail, about eight inches ; tail of the fame length : the female wants the falfe belly of the former ; but, on the lower part, the fkin forms on each fide a fold, betweeen which the teats are lodged.

This fpecies varies in color: I have feen one from *Guiana,* brown above, white beneath.

Inhabits the hot parts of *South America:* agrees with the others in its food, manners, and the prehenfile powers of its tail : it brings from ten to fourteen young at a time ; at left, in fome fpecies, there are that number of teats: the young affix themfelves to the teats as foon as they are born, and remain attached, like fo many inanimate things, 'till they attain growth and vigor to fhift a little for themfelves.

Cayopollin.

221. MEXICAN.

Cayopollin. *Hernandez Nov. Hisp.* 10.
Animal caudimanum. *Nieremberg*, 158.
Mus *Africanus* Kayopollin dictus, mas.
   Seb. Muf. tab. xxxi *fig.* 3.
Philander faturaté fpadiceus in dorfo, in
ventre ex albo flavicans, caudà ex fatu-
raté fpadiceo maculata. *Briffon quad.*
212. *Schreber*, cxlviii.
Le Cayopollin. *De Buffon*, x. 350. *tab.*
lv. LEV. MUS.

O. with large, angular, naked, and tranfparent ears: nofe
thicker than that of the former kind: whifkers very large:
a flight border of black furrounds the eyes: face of a dirty white,
with a dark line running down the middle: the hairs on the head,
and upper part of the body, afh-colored at the roots; of a deep
tawny brown at the tips: legs dufky: claws white: belly dull
cinereous: tail long, and pretty thick, varied with brown and
yellow: is hairy near an inch from its origin; the reft naked:
length, from nofe to tail, about nine inches; the tail the length
of the body and head.

Inhabits the mountains of *Mexico:* lives in trees, where it
brings forth its young: when in any fright, they embrace their
parent clofely: the tail is prehenfile, and ferves inftead of a
hand.

222. CAYENNE.

Le Crabier. *De Buffon, Supplem.* iii. 272.
Canis ferus major, *Cancrofus* vulgo dictus. *Koupara. Barrere France Æquinoct.* 149.

O. with a long flender face: ears erect, pointed, and fhort: the
coat woolly, mixed with very coarfe hairs, three inches
long, of a dirty white from the roots to the middle; from thence
to the ends of a deep brown: fides and belly of a pale yellow:
legs of a dufky brown: thumb on each foot diftinct: on the toes

of

of the fore feet, and thumb of the hind, are nails; on the toes of the hind feet crooked claws: tail very long, taper, naked, and fcaly.

Length feventeen *French* inches: of the tail fifteen and a half. The fubject meafured was young.

Inhabits *Cayenne:* very active in climbing trees, on which it lives the whole day. In marfhy places, feeds on crabs, which, when it cannot draw out of their holes with its feet, hooks them by means of its long tail. If the crab pinches its tail, the animal fets up a loud cry, which may be heard afar: its common voice is a grunt like a young pig. It is well furnifhed with teeth, and will defend itfelf ftoutly againft dogs: brings forth four or five young, which it fecures in fome hollow tree. The natives eat thefe animals, and fay their flefh refembles a hare. They are eafily tamed, and will then refufe no kind of food.

O. with the upper part of the head, and the back and fides, covered with long, foft, glofly hairs, of a dark cinereous color at the bottoms, and of a rufty brown towards the ends: belly of a dirty white.

223. New-Holland.

Tail taper, covered with fhort brown hairs, except for four inches and a half of the end, which was white, and naked underneath: toes like the former.

The fkin I examined had loft part of the face: the length from the head to the tail was thirteen inches: the tail the fame.

This was found near *Endeavour* river, on the eaftern coaft of

New Holland, with two young ones *. It lodges in the grafs, but is not common.

224. VULPINE·   *Stockdale's Bot. Bay,* 150.

O. with very long whifkers: ears erect, and pointed: upper parts of the body greyifh, mixed with dufky and white hairs, tinged with rufous: the laft predominates about the fhoulders: all the under fide of the neck and body of a tawny buff: about a quarter of the tail, next to the body, of the fame color with the back; the reft black: length from the tip of the nofe to the tail, two feet two inches: the tail fifteen.

     Inhabits *New Holland.*

225. SHORT-TAILED.   Mus fylveftris *Americana,* fœmina. *Seb.*   tre helvus, cauda brevi craffa, *Briffon Muf.* i. 50. *tab.* xxxi.   *quad.* 213. *Schreber,* cli.
Philander obfcurè rufus in dorfo, in ven-

O. with naked ears: the back of a dull red; belly of a paler: tail fcarce half the length of the body; thick at the bafe, leffening towards the end: no falfe belly.

     Inhabits *South America:* the young adhere to the teats as foon as born. *Seba* fays it lives in woods, and brings from nine to twelve young at a time.

              * *Cook's voy.* iii 586.

                                       Philander

Philander ex rufo luteus in dorfo, in ventre ex flavo albicans, capite craffo. *Briffon quad.* 213. *Seb. Muf.* i. 50. *tab.* xxxi.　*fig.* 8. *Klein quad.* 58.
Le Phalanger. *De Buffon,* xiii. 92. *tab.* x. xi. *Schreber,* clii.

226. PHALANGER.

O. with a thick nofe: fhort ears, covered with hair: eight cutting teeth in the upper jaw; two in the lower: hair on the upper part of the body reddifh, mixed with light afh-color, and yellow: the hind part of the head, and middle of the back, marked with a black line: the throat, belly, legs, and part of the tail, of a dirty yellowifh white; the reft of the tail brown and yellow: the body of the female marked with white: the firft and fecond toes of the hind feet clofely united: the claws large: the thumb on the hind feet diftinct, like that of the other fpecies: the bottom of the tail is covered with hair, for near two inches and a half; the reft naked: the length, from nofe to tail, near nine inches; the tail ten.

This fpecies inhabits the *Eaft Indian* iflands, as I am informed by Doctor *Pallas*; nor is it found in *Surinam,* as M. *de Buffon* conjectures.

PLACE.

De zak, of Beurs Rot. *Merian infect. Surinam.* 66. *tab.* lxvi.
Mus fylveftris Americana. *Seb. Muf.* i. 49. *tab.* xxxi. *fig.* 5.
Philander ex rufo helvus in dorfo, in ventre ex flavo albicans. *Briffon quad.* 212.

Mus fylveftris *Americanus,* catulos in dorfo gerens. *Klein quad.* 58.
Didelphis dorfigera. D. cauda bafi pilofa corpore longiore, digitis manuum muticis, *Lin. fyft.* 72.
Le Philandre de Surinam. *De Buffon,* xv. 157. Mus. Lev.

227. MERIAN *.

O. with long, fharp-pointed, naked ears: head, and upper part of the body, of a yellowifh brown color: the belly white,

* From *Merian,* a *German* paintrefs, who firft difcovered the fpecies at *Surinam.*

E 2

tinged

tinged with yellow: the fore feet divided into five fingers; the hind into four, and a thumb, each furnished with flat nails: tail very long, slender, and, except at the base, quite naked.

SIZE.

The length, from nose to tail, is ten inches. The tail exceeds the length of the body and head.

Inhabits *Surinam:* burrows under ground: brings five or six young at a time, which follow their parent: on any apprehension of danger, they all jump on her back, and twisting their tails round her's, she immediately runs with them into her hole.

* Flying.

228. FLYING.

Flying Opossum. *Stockdale's Bot. Bay,* 297. *White's,* 288.

O. with large ears: whole upper part of the body covered with a rich fur of a glossy black, mixed with grey. On each hip is a tan-color'd spot; all the under side white: tail at the base light color'd; increasing to black as it advances towards the tip: along the middle of the back from the head to tail, is a black line: on the fore feet are five toes; on the hind only three, with a thumb without any nail. From the fore to the hind feet, is a large membrane like the flying squirrel's: length from nose to tail, twenty inches: of the tail twenty two.

Inhabits *New Holland.* The fur exquisitely fine.

Kanguroo.

*Kanguru*___. *V.* 229.

** Gerboid.

Kanguroo. *Cook's voy.* iii. 577. *tab.* xx.
Yerboa gigantea. *Zimmerman,* 526.

O. with à fmall head, neck, and fhoulders: body increafing in thicknefs to the rump.

The head oblong, formed like that of a fawn, and tapering from the eyes to the nofe: end of the nofe naked and black: upper lip divided.

Noftrils wide and open: lower jaw fhorter than the upper: aperture of the mouth fmall: whifkers on both jaws: thofe on the upper longeft: ftrong hairs above and below the eyes.

Eyes not large; irides dufky; pupil of a blueifh black.

Ears erect, oblongly ovated, rounded at the ends, and thin, covered with fhort hairs; four inches long.

No canine teeth: four broad cutting teeth in the upper jaw: two long lanceolated teeth in the lower, pointing forward: four grinding teeth in each jaw, remote from the others. This animal has the very fingular power of feparating the lower incifores, and of bringing them again clofe to each other.

Belly convex and great.

Fore legs very fhort, fcarcely reaching to the nofe; ufelefs for walking.

Hind legs almoft as long as the body: the thighs very thick: on the fore feet are five toes, with long conic and ftrong claws; on the hind feet only three: the middle toe very long and thick, like that of an oftrich; and extends far beyond the two others,

TEETH.

LEGS.

5

which

which are placed very diftinct from it, and are fmall : the claws
fhort, thick, and blunt : the inner toe of the hind feet is fingu-
larly diftinguifhed by having on it two fmall claws : the bottom
of the feet, and hind part, black, naked, and tuberculated, as
the animal refts often on them.

TAIL.

Tail very long, extending as far as the ears; thick at the bafe,
tapering to a point : the end is black ; at the extremity is a ftrong
hard nail : the hair on all parts fhort and rather hard.

Scrotum large and pendulous, and is placed before the penis.

The female has on the belly an oblong pouch of a vaft depth.
The receptacle of its young.

Hair on the whole animal foft, and of an afh-color ; lighteft
on the lower parts.

SIZE.

Length of the largeft fkin I examined, three feet three inches
from the nofe to the tail : of the tail, two feet nine.

Weight of the largeft which was fhot, was eighty four pounds;
but this, on examination of the grinding teeth, had not attained
its full growth *. Later accounts inform us, they grow to the
weight of a hundred and forty pounds : to the length of fix
feet to the bafe of the tail : the tail itfelf, according to Mr. *Stockdale*'s publication, only two feet one.

PLACE.

Inhabits the weftern fide of *New Holland*, and has as yet been
difcovered in no other part of the world. The natives call it
Kangaru. It lurks among the grafs : feeds on vegetables : drinks
by lapping : goes chiefly on its hind legs, making ufe of the fore
feet only for digging, or bringing its food to its mouth. The
dung is like that of a deer. It is very timid : at the fight of men

---

* *Cook's voy*. iii. 586.

flies

flies from them by amazing leaps, fpringing over bufhes feven or eight feet high; and going progreffively from rock to rock. It carries its tail quite at right angles with its body when it is in motion; and when it alights often looks back: is much too fwift for gre-hounds: is very good eating, according to our firft navigators; but the old ones, according to the report of the more recent voygers, were lean, coarfe, and tough.

The weapon of defence was its tail, with which it would beat away the ftrongeft dog.

In the fpring of the prefent year I had opportunity of obferving the manners of one brought into the capital alive. It was in full health, very active, and very mild and good natured: on firft coming out of its place of confinement, it for a little time went on all fours, but foon affumed an upright attitude. It would fport with its keeper in a very fingular manner: it firft placed its tail in a perpendicular manner, erected its body on it as on a prop, and then raifing its whole body, darted its hind legs on the breaft of the man. It was capable of ftriking with great force if provoked: and it could fcratch violently with its fore claws.

This is a very anomalous animal: but has more relation to this genus than any other; and in form of its legs comes very near to the *Javan*. N° 219.

Kanguroo

O P O S S U M.

230. LESSER
KANGARU.

Kanguroo rat. *Stockdale, 277. White, 286.*

O. with the vifage of a rat; with two fharp pointed cutting teeth in the upper; two larger in the lower, with truncated ends: fore feet very fhort, furnifhed with four toes: hind legs and tail refembling the great fpecies. Three toes on each hind foot; the middle greatly exceeding the other two in length: on the belly is a pouch; within which were four nipples. The color above is of a pale brown, lighter on the belly: in fize double to that of the *Brown rat.*

MANNERS.

From the form of its parts, the manners probably the fame with thofe of the former: one was fhewn in *London* in 1790, but fo fhy as to elude a perfect defcription, continually concealing itfelf in the ftraw of the box.

231. SPOTTED
KANGARU.

Stockdale's Bot. Bay, 147.

O. with a long canine vifage: upright fharp ears: head and body black; the firft plain: the body and thighs marked with large fpots of white, thinly difperfed: tail covered with fhort hairs at the bafe; the reft very bufhy, covered with very long black hairs. Fore legs covered with fhort hairs for a fmall fpace next to the body; the remaining part naked: the feet furnifhed with five toes; the hind feet with four and a thumb, with a claw. Length from the nofe to the tail twenty-five inches: tail about nine.

Inhabits *New Holland.*

Six cutting teeth, two canine teeth, in each jaw.
Sharp nofe: flender bodies.
Five toes before; five behind.

232. COMMON.

Muftela. *Agricola An. Subter.* 485. *Gef-ner quad.* 752.
Weafel or Weefel, muftela vulgaris; in *Yorkfhire*, the Fitchet, or Fou-mart. *Raii fyn. quad.* 195.
The Whitred. *Sib. Scot.* iii. 11.
Wiefel. *Klein quad.* 62.
Muftela nivalis. *Lin. fyft.* 69.

Sno-mus. *Faun. Suec.* N° 18.
Muftela fupra rutila, infra alba. *Brif-fon quad.* 173.
La Belette. *De Buffon,* vii. 225. *tab.* xxix.
Weefel. *Br. Zool.* i. 82. *Br. Zool. il-luftr. tab.* ci. *Schreber,* cxxxviii.
LEV. MUS.

W with fmall rounded ears: whole upper part of the head and body of a pale tawny brown; under fide entirely white: a brown fpot beneath the corners of the mouth: length, from nofe to tail, between fix and feven inches: tail two and a half.

Inhabits moft parts of *Europe*; is common in *Siberia*, as far as *Kamtfchatka*; is met with in *N. America*, even as high as *Hudfon's Bay*; found alfo in *Barbary**. Is very deftruclive to chickens, birds, and young rabbits; a great devourer of eggs: does not eat its prey on the fpot; but after killing it, by a bite near the head, carries it off to its retreat: is a great deftroyer of field mice; a gentleman informed me he found eighty-five, new-ly killed, in one hole, which he believed belonged to this animal:

* *Shaw's Travels,* 249.

very active, runs up the sides of walls with great ease; no place is secure from its ravages: frequents outhouses, barns, and granaries: is a great enemy to rats and mice, and soon clears its haunts from those pernicious animals: brings four or five young at a time: its skin and excrements intolerably fœtid. In *Norway, Sweden, Russia,* and *Siberia,* it always changes to white at approach of winter. In *Siberia* it is called *Lasmitska:* their skins are sold to the *Chinese* for three or four rubles the hundred.

233. TOUAN. Le Touan *de la Cepedes,* &c. vi. 252. *tab.* lxi.

W with the upper part of head and body blackish; sides of the body, head, and legs, of a bright ferruginous: the lower part of the neck and body of a more pure white: the length from the nose to tail is rather more than five inches: the tail is rather more than two inches long, and tapers to a point.

Inhabits *Cayenne:* lives in hollow trees: lives on worms and insects, and brings two young at a time, which it carries on its back.

Mustela.

Muſtela. *Geſner quad.* 753.
Wieſel. *Kramer Auſtr.* 312. *Meyer's An.*
   ii. *tab.* 23, 24.
Muſtela erminea. M. plantis fiſſis;
   caudæ apice albo. *Lin. ſyſt.* 68.
Weſla. *Faun. ſuec.* No. 17.

β. ERMINE, when white. Mus Ponticus. *Plinii lib.* viii. *c.* 37. *Agricola An. Subter.* 484.
Armelinus, Hermelein. *Geſner quad.*
   754.
Gornoſtay. *Rzaczinſki Polen.* 235.
Muſtela candida, animal ermineum.

Muſtela hyeme alba, æſtate ſupra ru-
   tila infra alba, caudæ apice nigro.
   Briſſon quad. 176.
Le Roſelet. *De Buffon,* vii. 240. *tab.*
   xxix. *Schreber,* cxxxvii. A.
Stoat. *Br. Zool.* i. 84. LEV. MUS.

Raii ſyn. quad. 198.
L'Hermine. *De Buffon,* vii. 240. *tab.*
   xxix. *fig.* 2. *Briſſon quad.* 176. *Schre-
   ber,* cxxxvii. B.
Ermine. *Hiſt. Kamtſchatka,* 99. *Pontop.
   Norway.* ii. 25. *Br. Zool.* i. 84.
   LEV. MUS.

234. STOAT.

W with the upper part of the body pale tawny brown:
W. edges of the ears, and ends of the toes, of a yellowiſh
white: throat, breaſt, and belly, white: end of the tail black:
length, from noſe to tail, ten inches; tail five and a half: in
the N. of *Europe* and *Aſia,* and in the Highlands of *Scotland,* it
becomes entirely white at the approach of winter, the end of
the tail excepted: reſumes its brown color in the ſpring: ſometimes found white in *England:* one was brought to me in a
former winter, mottled with brown and white, the ſeaſon not
having been ſevere enough to effect a total change *; but in
February 1780, I ſaw in my grounds two others in the ſtate
of moſt perfect and beautiful *ermines.* In the mountains of
Southern *Aſia* and *Perſia,* it retains its brown color the whole
year†.

* *Br. Zool. illuſtr. tab.* ci.   † *Pallas.*

F 2   Inhabits,

Inhabits, in great abundance, the N. of *Europe,* and of *Afia;* in *Kamtfchatka* and the *Kurile* Iflands: is met with in *Newfoundland* and *Canada* *: the fkins a great article of commerce in *Norway* and *Siberia:* is found in the laft place in plenty, in birch forefts, but none in thofe of fir or pine: the fkins are fold on the fpot, from two to three pounds *fterling* per hundred †: taken in *Norway* in traps, baited with flefh; in *Siberia* ‡, either fhot with blunt arrows, or taken in a trap made of two flat ftones, propped by a ftick, to which is faftened a baited ftring, which, on the left touch of the animal, falls down and kills it; its manners and food the fame with the former; but does not frequent houfes: its haunts are woods and hedges, efpecially fuch as border on fome brook.

235. Quiqui.   Muftela Quiqui. *Molina Chili.* 273.

W. with a cuneiform nofe: ears fhort and round, with a white fpot in the middle: general color brown: legs and tail fhort: feet like thofe of a lizard: length from nofe to tail thirteen inches.

Inhabits *Chili:* is fierce and irritable: lives under ground: feeds on mice.

236. Cuja.   Muftela Cuja. *Molina Chili.* 272.

W. with black eyes: nofe turned up at the end: hair black; very thick, but foft: tail as long as the body, well furnifhed with hair: very like the ferret in fize, fhape, and teeth.

* *Charlevoix hift. Nouv. France,* v. 197.   † *Muller Ruff. Samlung.* 516.
‡ *Bell's travels,* i. 199. *Pontop. Norway,* ii. 25.

Inhabits *Chili*: lives on mice: breeds twice a year, and brings three or four at a time.

La Fouine de la Guiane. *De Buffon,* Suppl. iii. 161. *tab.* xxiii.

W. with a long fharp nofe: that, the cheeks, throat, and fides of the neck, black: forehead and fides of the head, to the ears, white: ears fhort, round, and edged with white: from each ear, a narrow white ftripe extends along the fides of the neck: the body covered with coarfe hairs, grey at their bafes, black and white at the ends: legs and feet black, tinged with red: the toes not unlike thofe of a rat.

Length from nofe to tail near twenty-one inches and a half: tail full of hair, of a bright chefnut, mixed with white; is rather fhorter in proportion than that of the *Englifh* Fitchet, to which it has a great refemblance.

Inhabits *Guiana*.

Putorius. *Gefner quad.* 767.
Yltis. *Agricola An. Subter.* 485.
Pole-cat, or Fitchet. *Raii fyn. quad.* 196.
Tchorz. *Rzaczinfki Polon.* 236.
Muftela fœtida. Iltis. Teuffels kind. *Klein quad.*
Muftela putorius. M. pedibus fiffis, corpore flavo nigricante; ore auri-culifque albis. *Lin. fyft.* 67. Iller. *Faun. fuec.* No. 16.
Muftela pilis in exortu ex cinereo albidis, colore nigricante terminatis, oris circumferentia alba. *Briffon quad.* 186.
Le Putois. *De Buffon,* vii. 199. *tab.* xxiii. *Schreber,* cxxxi.
Pole-cat. *Br. Zool.* i. 77. MUS. LEV.

W. with the fpace round the mouth white; the tips of the ears of the fame color: head, body, and legs, of a chocolate-color,

color, almoft black: on the fides the hairs are of a tawny caft: tail black: length feventeen inches; tail fix.

Inhabits moft parts of *Europe*; is common in the temperate parts of *Ruffia*, but grows fcarcer in *Siberia*, except in the defert of *Baraba*, and beyond the lake *Baikal*. None are found north of thofe places: they are ufually met with, in the places juft cited, with white or yellowifh rumps, bounded with black.

The Fitchet burrows under ground, forming a fhallow retreat, about two yards in length, generally terminating under the roots of fome large tree; fometimes forms its lodge under hayricks, and in barns: brings five or fix young at a time: preys on poultry, game, and rabbets: in winter frequents houfes, and will rob the dairy of milk. This animal is exceffively fœtid; yet the fkin is dreffed with the hair on, and ufed as other furs, for tippets, &c.; and is alfo fent abroad to line cloaths.

239. SARMATIAN. Muftela farmatica, *Ruffis* Perugufna. *Pallas, Itin.* i. 453. *Gueldenftaedt,* in Nov. Com. Petrop. xiv. 441. *tab.* x. *Zimmerman,* 486. *Schreber,* cxxxii. *Przewiafka,* or the girdled weefel? *Rzaczinfki,* auct. hift. *Polon.* 328.

W. with broad, fhort, round ears, edged with long white hairs: mouth furrounded with white: head, feet, and under fide of the body, of a full black: head croffed beyond each eye with a white band, paffing beneath the ears along the fides of the neck, and down to the throat: from the hind part of the head, another of yellow paffes on each fide obliquely towards

the

the fhoulders; above, is a third: the upper part of the body is of a brownifh black, ftriped and fpotted irregularly with obfcure yellow: tail dufky, full of hairs, intermixed with white ones longer than the reft: the end wholly black.

Length, from the tip of the nofe, about fourteen inches; of the tail fix.

SIZE.

Inhabits only *Poland*, and the fouthern provinces of *Ruffia*, between the *Dnieper* and *Volga*; and in *Afia*, the *Caucafean* mountains, and *Georgia*; and by report, *Bucharia*.

PLACE.

It is a moft voracious animal, feeding on the marmots, mice, and other leffer animals that inhabit with it the vaft plains of the *Ruffian* empire. Seizes on its prey, and firft fucks out the blood; does not meddle with eggs: lives ufually in holes made by other beafts, but is not without the power of burrowing: preys by night: fleeps little: very fierce and untameable: its eyes flaming: its fmell fœtid, efpecially when it erects its tail, which it does in anger: is very active: it moves by frequent jumps: copulates in the fpring: goes two months, and brings four or eight young, according to the report of the natives.

MANNERS.

Muftela Siberica, Kolonnok, *Ruffis. Pallas* Itin. 701.

240. SIBERIAN.

W. with the face black, whitifh about the noftrils, and fpotted towards the eyes; the reft of the animal of a deep yellow, nearly approaching to fox or orange color; with the throat fometimes fpotted with white: tail very full of hair, and of a

deeper

deeper color than the body: hair in general loofe and long: the foles of the feet thickly covered with fur.

SIZE.

Its body more flender than the Fitchet, coming nearer to the form of the Stoat: length to the tail twelve inches; of the tail fix.

PLACE.

Begins to appear in the *Altaic* mountains, between the *Ob* and the *Irtifh*, from whence it is common, in wooded mountains, to the *Amur* and lake *Baikal*. It has great refemblance in its manners, haunts, and food with the *fable*; but does not extend fo far north.

241. FERRET.

Viverra. *Plinii lib.* viii. *c.* 55. *Agricola An. Subter.* 486.
Muftela ruftica, viverra, Furo, Ictis. *Gefner quad.* 762. *Raii fyn. quad.* 198.
Fret. *Klein quad.* 63. *Schreber,* cxxxiii.
Viverra pilis fubflavis, longioribus, cafta-neo colore terminatis (mafc.) M. pilis ex albo fubflavis veftita. (fœm.) *Briffon quad.* 177.
Muftela Furo. M. pedibus fiffis, oculis rubicundis. *Lin. fyft.* 68. Mus. Lev.

W with a very fharp nofe: red and fiery eyes: round ears: color of the whole body a very pale yellow: length about fourteen inches; tail five.

Inhabits, in its wild ftate. *Africa** ; from whence it was originally brought into *Spain*†, in order to free that country from the multitudes of rabbets, with which the kingdom was overrun; from thence the reft of *Europe* was fupplied with it: is a lively active animal: the natural enemy of rabbets: fucks the

* *Shaw's travels,* 249.
† Καὶ γαλᾶς ἀϊρίας ἃς ἡ λυβύη φέρει *Strabo, lib.* iii.

blood

blood of its prey, feldom tears it: breeds in our climate: and brings five, fix, or nine at a time: but is apt to degenerate, and lofe its favage nature: warreners* are therefore obliged to procure an intercourfe between the female and a pole-cat, by leaving it near the haunts of the laft: the produce is a breed of a much darker color than the ferret, partaking more of that of the polecat. The ferret has the fame difagreeable fmell with that animal.

Martes gutture albo. *Agricola An. Subter.*
    485. *Gefner quad.* 764.
Stein-marter. *Klein quad.* 64.
Martes, alias Foyna, Martin, or Martlet.
    Raii fyn. quad. 200.
Kuna. *Rzaczinfki Polon.* 222.
Muftela pilis in exortu albidis caftaneo
    colore terminatis veftita, gutture albo.

Briffon quad. 178.
Muftela martes. M. pedibus fiffis, corpore fulvo nigricante, gula pallida. *Lin. fyft.* 67. Mard. *Faun. fuec.* No. 15.
La Fouine. *De Buffon,* vii. 186. *tab.* xviii.
    Schreber, cxxix.
Martin. *Br. Zool.* i. 79. LEV. MUS.

242. MARTIN.

W with broad rounded ears: lively eyes: head brown, • with a tinge of red: body, fides, and legs, covered with hair, afh-colored at the bottoms, bright chefnut in the middle, black at the tips: throat and breaft white: belly deep brown: tail full of hair, and of a dufky color: feet broad, covered at bottom with thick down: claws white: length eighteen inches: tail ten.

Inhabits moft parts of *Europe,* even to the warmer parts of *Ruffia,* but does not extend far eaft in that empire: is a moft elegant lively animal: capable of being tamed: is very goodnatured and fportive: lives in woods; and breeds in the hollow

PLACE.

* *Br. Zool.* i. 78. ii. 498.

of trees; and often, during winter, fhelters in magpies nefts:
brings from four to fix young at a time: deftroys poultry,
game, &c. and will eat rats, mice, and moles: the fkin and excrements have a mufky fmell: the fur is of fome value, and ufed
to line the robes of magiftrates.

**243. GREY-
HEADED.**

Le grande Marte de Guianne. *de La Cepedes. de Buffon,* Suppl. vi, 250. tab. lx.

W. with the head and upper part of the fides of the neck
greyifh: throat and under fide of the neck white: all
the reft of the body, limbs, and tail, black: length from the tip
of the nofe to the tail above two feet: of the tail (which is full of
hair) eighteen inches.

PLACE.   Inhabits *Guiana.*

244. PINE.

Martes gutture luteo. *Agricola An. Sub-
ter.* 485.
Martes fylveftris. *Gefner quad.* 765.
Martes abietum. *Raii fyn. quad.* 200.
Baum-Marter. *Klein quad.* 64.
Muftela pilis in exortu ex cinereo al-
bidis caftaneo colore terminatis, gut-
ture flavo. *Briffon quad.* 179.
La Marte. *De Buffon,* vii. 186. *tab.* xxii.
Schreber, cxxx.
Yellow-breafted Martin. *Br. Zool.* i. 81.
Faunul. Sinens. LEV. MUS.

W. with a yellow breaft and throat: the hair of a dark
chefnut-color, and of far fuperior finenefs to the former;
in other refpects agreeing with it.

                                                    Inhabits

Inhabits the N. of *Europe*, *Asia*, and *America*: found also in *Great Britain* *: are not found about the river *Oby*, nor in any part of *Siberia*: inhabits large forests, especially those of pines: never lodges near houses, as the other species is said † to do: M. *de Buffon* says, that it brings but two or three young at a time: its prey is the same with the former; its fur of far greater value. The peninsula of *Kamtschatka*, and *North America*, abound with them: their skins a prodigious article of commerce. Those found about Mount *Caucasus*, with an orange throat, are esteemed the finest in the furriers shops.

Zobela. *Agricola An. Subter.* 485.
Mustela Sobella. *Gesner quad.* 768.
Mustela Zibellina, the Sable. *Raii syn. quad.* 201. *Klein quad.* 64.
Mustela' Zibellina, *Aristotele* Satherius, *Nipho* Cebalus. *Alciato* Mus Samarticus et Scythicus. *Charleton Ex.* 20.
Mustela Zibellina. M. pedibus fissis, corpore obscurè fulvo, fronte exalbida, gutture cinereo. *Lin. syst.* 68.
Mustela Zibellina. *Nov. Com. Petrop.* v. 330. *tab.* vi.
Martes Zibellina. Mustela obscurè fulvo, gutture cinereo. *Brisson quad.* 180.
La Zibeline. *De Buffon*, xiii. 309.

245. SABLE.

W. with long whiskers: rounded ears: large feet: white claws: long and bushy tail; color of the hair black at the tips, cinereous at the bottom: chin cinereous, sometimes white, yellow, or spotted: the edges of the ears yellowish: sometimes the hair has a tawny cast; for in spring, after shedding

* M. *de Buffon* says, we have none of these animals in *England*, *Parce qu'il n'y a pas de bois*. That gentleman never did our kingdom the honour of making a progress through it.

† All foreign writers agree in this; but those which inhabit my neighborhood always keep in the woods, except in their nocturnal excursions.

the

the coat, the color varies: there are inftances of their being found of a fnowy whitenefs *.

The fize is equal to that of the Martin, to which it has a great refemblance in form: but this fpecific diftinction muft be noted— the tail of the *martin* is much longer than the hind legs, when extended : that of the Sable fhorter.

PLACE.   Inhabits *Siberia*, *Kamtfchatka*, and fome of the *Kurile* ifles, which lie between *Kamtfchatka* and *Japan*. Notwithftanding what Mr. *Scheffer* fays†, it is certain there are none to be found weft of the *Urallian* mountains, from whence they increafe in numbers, in proportion as you advance eaftward.

Sables live in holes in the earth, or beneath the roots of trees: fometimes, like the martin, form nefts in the trees, and will fkip with great agility from one to the other: are very lively, and much in motion during night: fleep much in the day: one that was kept tame would, on fight of a cat, fit up on its hind legs: excrements moft exceffively fœtid: prey, during fummer, on ermines, weefels, and fquirrels, but above all on hares; in winter, on birds; in autumn on hurtleberries, cranberries, and the berries of the fervice-tree: but during that feafon their fkins are at the worft, that diet caufing them to itch, and to rub off their fur againft the trees: they bring forth at the end of *March*, or beginning of *April*, and have from three to five at a time, which they fuckle for four or five weeks ‡.

Their chace was, in the more barbarous times of the *Ruffian* empire, the employ, or rather the tafks, of the unhapy exiles into *Siberia*: as that country is now become more populous, the

* *Strahlenberg hiſt. Ruſſia*, 442.   † *Scheffer Lapl.* 136.
‡ *Hiſt. Kamtfchatka*, 109, &c.

fables

fables have in great meafure quitted it, and retired farther *North* and *Eaft*, to live in defert forefts and mountains. They live near the banks of rivers, or in the little iflands in them *: on this account they have, by fome, been fuppofed to be the Σαθεριον of *Ariftotle*, *Hift. An. lib.* viii. c. 5; which he claffes with the animals converfant among waters.

At prefent the hunters of fables form themfelves into troops, from 5 to 40 each; the laft fubdivide into leffer parties, and each chufes a leader, but there is one that directs the whole: a fmall covered boat is provided for each party, loaden with pro-vifion, a dog and net for every two men, and a veffel to bake their bread in: each party alfo has an interpreter for the coun-try they penetrate into: every party then fets out according to the courfe their leader points out: they go againft the ftream of the rivers, drawing their boats up, till they arrive in the hunt-ing-country; there they ftop, build huts, and wait till the wa-ters are frozen, and the feafon commences. Before they begin the chace their leader affembles them, they unite in a prayer to the Almighty for fuccefs, and then feparate: the firft fable they take is called God's fable, and is dedicated to the church.

They then penetrate into the woods, mark the trees as they advance, that they may know their way back; and in their hunting-quarters form huts of trees, and bank up the fnow round them: near thefe lay their traps, then advance far-ther, and lay more traps, ftill building new huts in every quar-ter, and return fucceffively to every old one, to vifit the traps, and take out the game to fkin it, which none but the chief of

* *Avril's Travels,* 140.

the

the party muſt do: during this time they are ſupplied with proviſions by perſons who are employed to bring it on ſledges, from
the places on the road where they are obliged to form magazines, by reaſon of the impracticability of bringing quantities
thro' the rough country they muſt paſs. The traps are a ſort of
pit-fall, with a looſe board placed over it, baited with fiſh or
fleſh: when ſables grow ſcarce, the hunters trace them in the
new-fallen ſnow, to their holes, place their nets at the entrance,
and ſometimes wait, watching two or three days for the coming
out of the animal: it has happened, that theſe poor people have,
by the failure of their proviſions, been ſo pinched with hunger,
that, to prevent the cravings of appetite, they have been reduced
to take two thin boards, one of which they apply to the pit of
the ſtomach, the other to the back, drawing them tight together
by cords placed at the ends *: ſuch are the hardſhips our fellow-
creatures undergo, to ſupply the wantonneſs of luxury.

The ſeaſon of chace being finiſhed, the hunters re-aſſemble;
make a report to their leader of the number of ſables each has
taken; make complaints of offenders againſt their regulations;
puniſh delinquents; ſhare the booty; then continue at the head-
quarters 'till the rivers are clear of ice; return home and give
to every church the dedicated furs.

COMMERCIAL
HISTORY. The following is the commercial hiſtory of this fur-trade,
which Dr. *J. R. Foſter* was ſo obliging as to tranſlate for me,
from *Muller's Samlung Ruſs. Geſchichte,* iii. 495 to 515, being
an abſtract from above 20 pages.

* *Bell's Travels,* i. 245.

" SABLE,

" SABLE, SOBOL in *Ruſſian*; ZOBEL in *German:* their price
varies, from 1 l. to 10 l. ſterling, and above: fine and middling
ſable ſkins are without bellies, and the coarſe ones are with
them: forty ſkins make a collection called *Zimmer:* the fineſt
ſables are ſold in pairs, perfectly ſimilar, and are dearer than
ſingle ones of the ſame goodneſs; for the *Ruſſians* want thoſe in
pairs for facing caps, cloaks, tippets, &c. The blackeſt are re-
puted the beſt. Sables are in ſeaſon from *November* to *February*;
for thoſe caught at any other time of the year are ſhort haired,
and then called *Nedoſoboli.* The hair of ſables differs in length and
quality: the long hairs, which reach far beyond the inferior
ones, are called *Os*; the more a ſkin has of ſuch long hairs, the
blacker they are, and the more valuable is the fur; the very
beſt have no other but thoſe long and black hairs. *Motchka* is
a technical term in the *Ruſſian* fur-trade, expreſſing the lower
part of the long hairs; and ſometimes it comprehends likewiſe
the lower and ſhorter hairs: the above mentioned beſt ſable
furs are ſaid to have a black *Motchka.* Below the long hairs are, in
the greater parts of ſable furs, ſome ſhorter hairs, called *Podoſie,*
i. e. *Under-Os:* the more *Podoſie* a fur has, the leſs valuable: in
the better kind of ſables the *Podoſie* has black tips, and a grey or
ruſty *Motchka:* the firſt kind of *Motchka* makes the middling
kind of ſable furs; the red one the worſt, eſpecially if it has but
few *Os:* between the *Os* and *Podoſie* is a low woolly kind of hair,
called *Podſada*; the more *Podſada* a fur has, the leſs valuable,
for the long hair will, in ſuch caſe, take no other direction than
the natural one; for the character of ſables is, that notwithſtanding the hair naturally lies from the head towards the tail, yet will

it lie equally in any direction, as you ftrike your hand over it: the various combinations of thefe characters, in regard to *Os*, *Motchka*, *Podofie*, and *Podfada*, make many fpecial divifions of the goodnefs of furs: befides this, the furriers attend to the fize, preferring always, *cæteris paribus*, the biggeft, and thofe that have the greateft glofs: the fize depends upon the animal being a male or female, the latter being always fmaller. The glofs vanifhes in old furs: the frefh ones have a kind of *bloomy* appearance, as they exprefs it; the old ones are faid to have done *blooming*: the dyed fables always lofe their glofs, become lefs uniform, whether the lower hairs have taken the dye or not, and commonly the hairs are fomewhat twifted or crifped, and not fo ftrait as in the natural ones: fome fumigate the fkins, to make them look blacker; but the fmell, and the crifped condition of the long hair, betrays the cheat; and both ways are detected, by rubbing the fur with a moift linen cloth, which grows black in fuch cafes.

" The *Chinefe* have a way of dying the fables, fo that the color not only lafts, (which the *Ruffian* cheats cannot do) but the fur keeps its glofs, and the crifped hairs only difcover it: this is the reafon that all the fables, which are of the beft kind, either in pairs or feparate, are carried to *Ruffia*; the reft go to *China*: the very beft fables come from the environs of *Nertchifk* and *Yakutfk*; and in this latter diftrict, the country about the river *Ud* affords fometimes fables, of whom one fingle fur is often fold at the rate of 60 or 70 rubles (12 or 14 l.) The *bellies* of fables, which are fold in pairs, are about two fingers breadth, and are tied together by forty pieces, which are fold from 1 to 2 l. fterling. *Tails*
are

are fold by the hundred; the very beft fable furs muft have their tails, but ordinary fables are often cropped, and a hundred fold from 4 to 8 l. fterling: the *legs* or *feet* of fables are feldom fold feparately. White fables are rare, and no common merchandize, but bought only as curiofities: fome are yellowifh, and are bleached in the fpring on the fnow."

The common fables are fcarcely better in hair and color than the martin.

The fable is found again in *North America.* The *Ruffians* have often difcovered the fkins mixed with thofe of martins, in the fur-dreffes which the *Ichutcki* get from the *Americans* by way of exchange. Their fur is more gloffy than that of the *Siberian* fable, and of a bright chefnut-color; but of a coarfer quality. It is to be obferved, that no fables are found N. E. of the river *Anadyr*, the country of the *Ichutcki**.

The information I received from Doctor *Pallas,* refpecting the character of this animal, obliges me to lay afide my notion of its being found in the new world, under the name of *The Fifher*; yet I have reafon to fuppofe I have recovered it on that continent, by feeing the fkin of another quadruped highly refembling it, in the cabinet of Mrs. *Blackburn,* fent from *Canada*; which I defcribe under the name of the *American.*

Its length, from nofe to tail, was twenty inches. The trunk of the tail only five inches: but from the rump to the end of the hairs eight. The ears more pointed than thofe of the *Afiatic* fable: feet very large, hairy above and below: five toes, with white claws on each foot.

AMERICAN.

* Doctor *Pallas.*

COLOR.

Color of the head and ears whitifh: whifkers fhort and black: whole body of a light tawny: feet brown. This feems to have been one of the bleached kind before mentioned.

246. FISHER.

W. with a black nofe: ftrong and ftiff whifkers: fix fmall weefel-like teeth above and below: fix large canine teeth: four grinding teeth in each upper jaw; three fharp-pointed, the fourth flat: in the lower jaws fix; the laft flauted, the next tridentated; the next to thofe bidentated: ears broad and round, dufky on their outfides, edged with white: face and fides of the neck pale brown, or cinereous, mixed with black: hairs on the back, belly, legs, and tail, black; brownifh at their bafe: fides brown: the feet very broad; covered with hair even on their foles: five toes on the fore feet; generally four, but fometimes five, on the hind feet; with fharp, ftrong, and crooked white claws: fore. legs fhorter than thofe behind: tail full and bufhy, fmalleft at the end, feventeen inches long: length, from nofe to tail, twenty-eight inches.

Inhabits *North America:* notwithftanding its name, is not amphibious: preys on all forts of leffer quadrupeds*: by the number of fkins imported, is not an uncommon animal; not lefs than 580 being brought in one feafon from *New York* and *Penfylvania:* feems to be the animal called by *Joffelyn*†, the SABLE; which, he fays, is perfectly black. I have feen many of the fkins, which vary in color. LEV. MUS.

* By a letter from Mr. *Peter Collinfon,* who received the account from *Bartram,* of *Penfylvania.*

† *Joffelyn's voy.* 87.

Le

Le Vanfire. *De Buffon*, xiii. 167. *tab.* xxi. de la *Cepedes, de Buffon*, Suppl, vii. 249. tab. lix.

W with fhort ears: the hair on the whole body brown at the roots, and barred above with black, and ferruginous: the tail of the fame color: the length from nofe to tail, about fourteen inches; the tail, to the tip of the hairs at the end, near ten.
Inhabits *Madagafcar.*

Le Pekan. *De Buffon*, xiii. 304. *tab.* xlii. *Schreber,* cxxxiv. LEV. MUS.

W with very long and ftrong whifkers: ears a little pointed: hair on the head, back, and belly, cinereous at the roots, of a bright bay at the ends; very foft and gloffy: on the fides is a tinge of grey: between the fore-legs a white fpot: legs and tail black: toes covered with thick hair, above and below: claws fharp.
In form like a martin: its length, from nofe to tail, one foot feven inches: the length of the trunk of the tail above ten; and the hairs extend an inch beyond.
Inhabits *North America:* defcribed from a fkin.

Le Vifon. *De Buffon,* xiii. 308. *tab.* xliii.

W with rounded ears: color of the hair brown, tinged with tawny, very bright and gloffy: beneath is a thick down, cinereous tipt with ruft color: legs very fhort: tail dufky.

H 2 Length

SIZE.   Length to the tail above feventeen inches: tail, to the extremity of the hairs, nine.

PLACE.   Inhabits *North America*: defcribed from the ftuffed fkins, in 1765, in the cabinet of M. *Aubry*, curate of *Saint Louis*, in *Paris*. A fuller account of this and the preceding animal is defired.

250. WHITE-CHEEKED.   W. with rounded ears: broad and blunt nofe: dufky irides: head flat: face, crown, legs, rump, and tail, black: chin and cheeks white: throat of a rich yellow: back and belly of a pale yellow, intimately mixed with cinereous.

SIZE.   Length, from nofe to tail, eighteen inches: tail of the fame length, covered with long hair.

Defcribed from the living animal at Mr. *Brooks*'s, *April* 1774. Place unknown.

251. GRISON.   Le Grifon. *De Buffon*, xvi. 169. *tab.* xxv. *Allamand*, v. 65. *tab.* viii. *Schreber*, cxxiv.

W. with large head and eyes: fhort but broad ears: upper part of the body of a deep brown, each hair tipped with white, which gives it a grey or hoary look: from each fide of the forehead extends a broad white line, paffing over the eyes, and reaching as far as the fhoulders: the nofe, throat, and whole under fide of the body, thighs, and legs, black.

SIZE.   Length, from the tip of the nofe to the origin of the tail, feven inches. By the figure, the tail is little more than half the length of the body.

Inhabits

Inhabits *Surinam*, but is a very fcarce animal: firft defcribed by Mr. *Allamand.*

Galera, fubfufca, cauda elongata, auribus fubnudis appreffis. *Browne's Jamaica,* 485. *tab.* xlix.

Le Tayra, ou le Galera. *De Buffon,* xv. 155. *Schreber,* cxxxv.

252. GUINEA.

W • with the upper jaw much longer than the lower: eyes placed mid-way between the ears and tip of the nofe: ears like the human: tongue rough: tail declining downwards, leffening towards the point: feet ftrong, and formed for digging: fhape of the body like that of a rat: fize of a fmall rabbet: of a dufky color: the hair rough.

Inhabits *Guinea:* common about the negro fettlements: burrows like a rabbet: very fierce; if drove to neceffity will fly at man or beaft: very deftructive to poultry: feems to be the *Kokeboe* of *Bofman* *, which only differs in color, being red.

PLACE.

Muftela barbara. M. pedibus fiffis, atra, collo fubtus macula alba triloba. *Lin. fyft.* 67.

Muftela maxima atra mofcum redolens. *Tayra,* groffe Belette. *Barrere France Æquin.* 155.

253. GUIANA.

W • with round ears, covered with down: an afh-colored fpace between the eyes: a trilobated fpot on the lower part of the neck: fize of a martin: color black: hairs coarfe.

Inhabits *Brafil* and *Guiana:* when it rubs itfelf againft trees, leaves behind an unctuous matter, that fcents of mufk.

PLACE.

* *Hift. Guinea,* 239.

La

254. WOOLLY.　　　La petite Fouine de la Guiane. *De Buffon, Suppl.* iii. 162. *tab.* xxiv.

W with a long slender nose: upper jaw longer than the
· lower: ears very short and round: body covered with
woolly hair: tail taper, ending in a point, between eight and nine
inches long: body, from nose to tail, between fifteen and sixteen.

M. *de Buffon* does not mention the color; by his figure the
belly seems white. He says it inhabits *Guiana.* I am doubtful
whether it is not one of the above species.

255. ICHNEUMON.

Ιχνευμων. *Ariſtot. hiſt. An. lib.* ix. *c.* 6.
　　Oppian Cyneg. iii. 407.
Ichneumon. *Plinii lib.* viii. *c.* 24.
L'Ichreumon, que les *Egyptiens* nom-
　　ment Rat *.de Pharaon. *Belon obſ.* 95.
　　Portraits, 06. *Proſp. Alp.* i. 234. *Geſ-
　　ner quad.* 566. *Raii ſyn. quad.* 202.
　　Shaw's Travels, 249, 376.
Muſtela Ægyptiaca. *Klein quad.* 64.

β. INDIAN. Quil, vel Quirpele. *Garcia
　　Arom.* 214. *Raii ſyn. quad.* 197.
Viverra Mungo. *Kæmpfer Amœn.* 574.
De Mongkos. *Valentyn Amboin.* iii.
Serpenticida five Moncus. *Rumph. herb.
　　Amboin. App.* 69. *tab.* xxviii.
Indian Ichneumon. *Edw.* 199.
Ichneumon feu vulpecula Ceilonica.

Meles Ichneumon digitis mediis longio-
　　ribus, lateralibus æqualibus, unguibus
　　ſubuniformibus. *Haſſelquiſt itin.* 191.
Ichneumon: Mus *Pharaonis* vulgo. *Briſ-
　　ſon quad.* 181.
Viverra Ichneumon. V. cauda e baſi in-
　　craſſata ſenſim attenuata, pollicibus re-
　　motiuſculis, *Lin. ſyſt.* 63. *Schreber,* cxvi.
　　A. LEV. MUS.

Seb. *Muſ.* i 66. *tab.* xli. *fig.* 1.
La Mangouſte. *De Buffon,* xiii. 150. *tab.*
　　xix. Le Nems, *tom.* xvi. 104 *tab.*
　　xxvii.
Viverra indica. V. ex griſeo rufeſcens.
　　Briſſon quad. 177. *Raii ſyn. quad.* 198.
　　Schreber, cxvi. LEV. MUS.

W with bright flame-colored eyes: small rounded ears, al-
· moſt naked: noſe long and ſlender: body thicker than

* The *Ægyptians* never ſtyle it *Phar,* or Mouſe, but *Nems,* or Ferret, from its
reſemblance to that animal. *Haſſelquiſt,* 196. This *Forſkal* confirms, p. 111.

　　　　　　　　　　　　　　　　　　　　　　　　　　　that

that of others of this genus: tail very thick at the bafe, tapering to a point: legs fhort: the hair is hard and coarfe: color various in different animals, from different countries; in fome alternately barred with dull yellowifh brown and white; in others, pale brown and moufe-coloured; fo that the animal appears mottled: throat and belly of a uniform brown: beneath the tail is an orifice not unlike that of a badger.

The fpecimen in the *Afhmolean Mufeum* was thirteen inches and a half long to the origin of the tail; the tail eleven: the *Ægyptian* variety is the largeft. Some are forty-two inches long from the nofe to the extremity of the tail. M. *de Buffon* gives the figure of one in the xxvith plate of his Supplement, vol. iii. under the name of *La Grande Mangoufte*: the tail is longer, and more flender than that of the common kind, and the hair univerfally more broken and coarfer.

Inhabits *Ægypt, Barbary, India,* and its *iflands:* a moft ufeful animal; being an inveterate enemy to the ferpents and other noxious reptiles which infeft the torrid zone: attacks without dread that moft fatal of ferpents the *Naja,* or *Cobra di Capello;* and fhould it receive a wound in the combat, inftantly retires; and is faid to obtain * an antidote from a certain herb; after

which

---

* A faCt, as yet, not well eftablifhed: Botanifts are not yet agreed about the fpecies of this fanative plant, whofe ufe, it is pretended, this weefel pointed out to mankind: thofe who have feen the combats between the *Ichneumon* and *Naia,* never could difcover it: *Kæmpfer,* a writer of the firft authority, who vifited *India,* and who had a tame Ichneumon, and been witnefs to its battles with the ferpent, fays no more than that it retired and eat the roots of any herb it met with. It is from the *Indians* he received the account. of the root, whofe veracity

he

which it returns to the attack, and feldom fails of victory. Is a
great deftroyer of the eggs of crocodiles, which it digs out of
the fand; and even kills multitudes of the young of thofe terrible
reptiles: it was not therefore without reafon, that the antient
Ægyptians ranked the *Ichneumon* among their Deities: is at prefent domefticated, and kept in houfes in *India* and in *Ægypt*; for
it is more ufeful than a cat, in deftroying rats and mice: grows
very tame: is very active; fprings with great agility on its prey;
will glide along the ground like a ferpent, and feems as if without feet: fits up like a fquirrel, and eats with its fore feet:
catches any thing that is flung to it: is a great enemy to poultry:
will feign itfelf dead till they come within reach: loves fifh:
draws its prey, after fucking the blood, to its hole: its excrements very fœtid: when it fleeps, brings its head and tail under
its belly, appearing like a round ball, with two legs fticking out.
Rumphius obferves how fkilfully it feizes the ferpents by the throat,
fo as to avoid receiving any injury: and *Lucan* beautifully de-

he fpeaks moft contemptuoufly of. *Amœn. Exot.* 576. *Rumphius* never faw the
plant growing; but defcribes it from a fpecimen fent him from *Java*; for he
fays the *Indians* would perfuade him that it had no leaves. Vide *Herb. Amboin.
App.* 71. All that feems certain is, that the *Indians* have a plant, of whofe
alexipharmic virtues they have a high opinion, and are faid to ufe it with fuccefs
againft the dreadful *macaffar* poifon, and the bite of ferpents. *Kæmpfer* fays he
had good fuccefs with one fpecies, in putrid fevers, and found it infallible for the
bite of a mad dog. As there is no doubt but a moft ufeful plant of this nature
does exift in the *Indies,* it is to be hoped that ftrict enquiry will be made after it.
In order to direct their fearches, they are referred to

Garcia ab Horto Hift. Aromatum in Clus. Exot. 214.

Kæmpfer Amœn. Exot. 573. *Rumph. Herb. Amboin. App.* 29.

Amœn. Acad. ii. 89. *Flora Zeylanica,* 46. 190, 239.

                                                            fcribes

fcribes the fame addrefs of this animal, in conquering the *Ægyptian* Afp.

Afpidas ut Pharias *cauda folertior hoftis*
Ludit, et iratas incerta provocat umbra :
Obliquanfque caput vanas ferpentis in auras
Effufæ toto comprendit guttura morfu
Letiferam citra faniem : tunc irrita peftis
Exprimitur, faucefque fluunt pereunte veneno. Lib. iv. 724.

Thus oft' th' *Ichneumon*, on the banks of *Nile*,
Invades the deadly *Afpic* by a wile ;
While artfully his flender tail is play'd,
The ferpent darts upon the dancing fhade ;
Then turning on the foe with fwift furprize,
Full on the throat the nimble feizer flies :
The gaping fnake expires beneath the wound,
His gufhing jaws with poifonous floods abound,
And fhed the fruitlefs mifchief on the ground.

ROWE.

Gm. Lin. 85.   256. CAFRE.

W with fhort hairy ears : hairs on the body fhining, rude, mixed with yellow, black, and brown : tail grows gradually more flender from the bafe, tip black.
Inhabits the *Cape* of *Good Hope*.

Le Surikate. *De Buffon*, xiii. 72. *tab.* viii. *Schreber*, cxvii. *Miller's plates*, xx.   257. FOUR-TOED.

W with a very fharp-pointed nofe : head depreffed : cheeks inflated : upper jaw much longer than the lower ; tip black: whiſkers black, ariſing from warty tubera: irides duſky: region about the eyes black: ears ſmall, rounded, black, lying cloſe to the head.

Tongue oblong, blunt, aculeated backwards.

Six ſmall inciſores; two long canine in each jaw, and five grinders on each ſide.

Back very broad, and a little convex: belly broad and flat.

Legs ſhort: feet ſmall, naked at the bottom; four toes on each: the claws on the fore feet long, like thoſe of the badger; on the hind feet ſhort.

Color of the hairs brown near the bottom; black near the ends, and hoary at the points; thoſe on the back undulated: inſide of the legs yellowiſh brown: tail tufted with black.

Length from noſe to tail eleven inches; of tail eight: the laſt thick at the baſe, ending pretty abrupt.

Inhabits the *Cape of Good Hope*, where it is called *Meer-rat:* feeds on fleſh; preys on mice; is a great enemy to *Blattæ:* is always making a grunting noiſe: is much in motion: ſits quite erect, dropping its fore legs on its breaſt, and moving its head with great eaſe, as if on a pivot, and appearing as if it liſtened, or had juſt ſpied ſomething new. When pleaſed, it makes a rattling noiſe with its tail, for which reaſon the *Dutch* at the *Cape* call it *Klapper-maus**. It is alſo found in *Java*, where the *Javaneſe* ſtyle it *Jupe*; the *Dutch*, *Suracatje**. The animal which I examined was brought alive from the *Cape*. Well engraven in *Miller's* plates, *tab.* xx.

* *Pallas* Miſcel. Zool. 59, 60.

Yellow

Yellow Weasel __ N.° 258.

Yellow maucauco. *Syn. quad.* No. 108. Viverra caudivolvola. *Schreber,* tab. xlii.

W with a fhort dufky nofe: fmall eyes: ears fhort, broad, and flapping, and placed at a great diftance from each other: head flat and broad: cheeks fwelling out: tongue very long: legs and thighs fhort, and very thick: five toes to each foot, feparated and ftanding all forward: claws large, a little hooked, and of a flefh-color.

The hairs fhort, foft, gloffy, clofely fet together: on the head, back, and fides a mixture of yellow and black: cheeks, infide of the legs, and the belly, yellow: half way down the middle of the belly is a broad dufky lift, ending at the tail; and another from the head along the middle of the back to the tail: tail of a bright tawny, mixed with black; is round, and has the fame prehenfile faculty as fome of the monkies have: length from the nofe to the tail nineteen inches; of the tail feventeen.

It was very good-natured and fportive; would catch hold of any thing with its tail, and fufpend itfelf: lay with its head under its legs and belly. — MANNERS.

Shewn about twelve years ago in *London:* its keeper faid it came from the mountains of *Jamaica,* and called it a *Potto,* the name given by fome writers to a fpecies of *Sloth* found in *Guinea.* LEV. MUS. — PLACE.

Le Kinkajou. *De Buffon,* xvi. 244. tab. l.

W with a fhort dufky nofe: tongue of a vaft length: fmall eyes, encircled with dufky: ears fhort and rounded, and placed very diftant: the hairs fhort; on the head, upper part of the body, and the tail, the colors are yellow, grey, and black intermixed: the fides of the throat, and under fide, and the infides of the legs, of a lively yellow: the belly of a dirty white, tinged with yellow.

The toes feparated: the claws crooked, white, guttered beneath.

The length from head to tail two feet five *(French)*; of the tail, one foot three: the tail is taper, covered with hair, except beneath, near the end, which is naked, and of a fine flefh-color. It is extremely like the former; but larger in all its parts.

Like the former, it has a prehenfile tail, and is naturally very good-natured: goes to fleep at approach of day; wakes towards night, and becomes very lively: makes ufe of its feet to catch at any thing: has many of the actions of a monkey: eats like a fquirrel, holding the food in its hands: has variety of cries during night; one like the low barking of a dog: its plantive note is cooing; its menacing, hiffing; its angry, confufed.

Is very fond of fugar, and all fweet things: eats fruits, and all kinds of vegetables: will fly at poultry, catch them under the wing, fuck the blood, and leave them without tearing them: prefers a duck to a pullet; yet hates the water.

M. *de Buffon* calls this animal *le Kinkajou,* after a defcription

(given

P Mazell Pecdp

Brasilian Weesel ____ N.° 260.

(given by M. *Dennis*) of one of that name found in *N. America,* described also by *Charlevoix,* under the name of *Carcajou*; both which, in fact, are the same as my *Puma,* N° 189. M. *Dennis* gives it the same manners; adds, that it climbs trees, watches the approach of the moose, falls on, and soon destroys it. He says, he lost a heifer by one of those animals, which at once eat through its neck; but the quadruped in question never could have the powers attributed to so ferocious a creature. This therefore is new, and by form and manners a proper concomitant of the animal last described.

This animal was brought to *Paris* from *New Spain,* and lived there two or three years. It is a very distinct species from the former, of which M. *de Buffon* gives a very indifferent figure, taken from the animal I describe.

Coati. *Marcgrave Brasil.* 228. *De Laet,* 486. *Raii syn. quad.* 180. *Klein quad.* 72.

Vulpes minor, rostro superiore longiusculo, cauda annulatim ex nigro et rufo variegatâ. Quachy. *Barrere France Æquin.* 167.

Viverra nasua. V. rufa, cauda albo annu- lata. *Lin. syst.* 64.

Ursus naso producto et mobili, cauda an- nulatim variegata. *Brisson quad.* 190.

Le Coati brun. *De Buffon,* viii. 358. *tab.* xlviii. *Schreber,* cxviii.

Badger of *Guiana. Bancroft,* 141. LEV. Mus.

260. BRASILIAN.

W with the upper jaw lengthened into a pliant, moveable • *proboscis,* much longer than the lower jaw: ears round- ed: eyes small: nose dusky: hair on the body smooth, soft, and glossy, of a bright bay color: tail annulated with dusky and bay: breast whitish: length, from nose to tail, eighteen inches; tail, thirteen.

4

β. DUSKY.

β. Dusky. Nose and ears formed like the preceding: above and beneath the eye two spots of white: hair on the back and sides dusky at the roots, black in the middle, and tipt with yellow: chin, throat, sides of the cheeks, and belly, yellowish: feet black; tail annulated with black and white; sometimes the tail is of an uniform dusky color *. *Le Coati noiatre* of M. *de Buffon, tab.* xlvii. *Schreber*, cxix. The *Coati-mondi* of *Marcgrave*.

Inhabits *Brasil* and *Guiana:* feeds on fruits, eggs, and poultry: runs up trees very nimbly: eats like a dog, holding its food between its fore-legs: is easily made tame: is very good-natured: makes a sort of whistling noise: seems much inclined to sleep in the day. *Marcgrave* observes, that they are very subject to gnaw their own tails.

261. Stifling.   Yzquiepatl. *Hernandez Mex.* 332. *Raii*   *tab.* xlii.
      syn. quad. 181. *Klein quad.* 72.   Le Coase. *De Buffon?* xiii. 288. *tab.*
      Meles Surinamensis. *Brisson quad.* 185.   xxxviii. *Schreber,* cxx.
      Ichneumon de Yzquiepatl. *Seb. Mus.* i.

W with a short slender nose: short ears and legs: black body, full of hair: tail long, of a black and white color: length, from nose to tail, about eighteen inches.

Inhabits *Mexico,* and perhaps other parts of *America.* This, and the four following species, remarkable for the pestiferous,

* Described as a distinct species by *Linnæus,* under the title of *viverra Narica.* V. *subfusca, cauda unicolore,* 64. and by M. *Brisson,* under that of *Ursus naso producto et mobili, cauda unicolore,* 190.

suffocating

fuffocating and moft fœtid vapour they emit from behind, when attacked, purfued, or frightened: it is their only means of defence: fome turn * their tail to their enemy, and keep them at a diftance by a frequent *crepitus*; and others ejaculate their urine, tainted with the horrid effluvia, to the diftance of eighteen feet: the purfuers are ftopped by the terrible ftench: fhould any of this liquid fall into the eyes, it almoft occafions blind-nefs; if on the cloaths, the fmell will remain for feveral days, in fpite of all wafhing; they muft even be buried in frefh foil, in order to be fweetened. Dogs that are not true bred, run back as foon as they perceive the fmell; thofe that have been ufed to it, will kill the animal; but are often obliged to relieve them-felves by thrufting their nofes into the ground. There is no bearing the company of a dog that has killed one, for feveral days.

Profeffor *Kalm* was one night in great danger of being fuffo-cated by one that was purfued into a houfe where he flept; and it affected the cattle fo, that they bellowed through pain. Ano-ther, which was killed by a maid-fervant in a cellar, fo affected her with its ftench, that fhe lay ill for feveral days: all the pro-vifions that were in the place were fo tainted, that the owner was obliged to throw them away.

Notwithftanding this, the flefh is reckoned good meat, and not unlike that of a pig: but it muft be fkinned as foon as kill-ed, and the bladder taken carefully out. The *Virginian* fpecies,

* *Wood's voy.* in *Dampier*, iv. 96; the reft of the account is taken from *Catef-by* and *Kalm*.

or *skunk*, is capable of being tamed, and will follow its master like a dog: it never emits its vapour, except terrified.

It breeds in hollow trees, or holes under ground, or in clefts of rocks: climbs trees with great agility: kills poultry, eats eggs, and destroys young birds.

262. STRIATED.

Pole-cat, or Skunk. *Lawson Carolina.*
Pole-cat. *Catesby Carolina,* ii.
Mustela Americana fœtida. *Klein quad.* 64.
Mustela nigra tæniis in dorso albis. *Brisson quad.* 181.

Viverra putorius. V. fusca lineis quatuor dorsalibus parallelis albis. *Lin. syst.* 64.
Le Conepate. *De Buffon,* xiii. 288. *tab.* xl. *Schreber,* cxxii.

W with rounded ears: head, neck, belly, legs, and tail, black: the back and sides marked with five parallel white lines: one on the top of the back; the others on each side: the second extends some way up the tail, which is long and bushy towards the end: size of an *European* Pole-cat; the back more arched: varies in the disposition of the stripes.

Inhabits *N. America:* when attacked, bristles up its hair, and flings its body into a round form: its vapour horrid. *Du Pratz* says, that the male of the *Pole-cat,* or *Skunk,* is of a shining black: perhaps the *Coase* of M. *de Buffon* is the male; for his description does not agree with the *Yzquiepatl,* which he makes synonymous.

Chinche.

Chinche. *Feuilleè obſ. Peru*, 1714, *p.* 272. *voix Nouv. France*, v. 196.
Skunk, Fiſkatta. *Kalm's voy. Forſter's* Le Chinche. *De Buffon*, xiii. 294. *tab.*
tr. i. 273. *tab.* ii. *Joſſelyn's voy.* 85. xxxix. *Schreber*, cxxi. Lev. Mus.
Enfant du Diable, Bete puante. *Charle-*

W with ſhort rounded ears: black cheeks: a white ſtripe
• from the noſe, between the ears, to the back: upper
part of the neck, and the whole back. white; divided at the bottom by a black line, commencing at the tail, and paſſing a little
way up the back: belly and legs black: tail very full of long
coarſe hair; generally black, ſometimes tipt with white, and
ſometimes wholly white *, that figured by M. *de Buffon* entirely
white: nails on all the feet very long, like thoſe on the fore-feet
of a badger. Rather leſs than the former.

Inhabits *Peru*, and *N. America*, as far as *Canada:* of the ſame
manners and ſtench with the others.

Viverra Cinghe. *Molina Chili.* 269.

W with black hair, changeable into blue: along the back a
• bed of white round ſpots from head to tail: head long:
ears large, well covered with hair, and pendulous: hind legs longer than the fore.

Inhabits *Chili:* carries its head low: back arched; which it

* *De la Cepidès de Buffon,* Suppl. *tom.* vii. p. 233. *tab.* lvii.

MANNERS.

generally covers with its bufhy tail, like the fquirrel: digs holes
in the ground, in which it hides its young.

In manners and food agrees with the *Stifling*; and its dreadful
ftench. *Molina* denies that the fmell comes from the urine, but
from a greenifh oil coming from a bladder feated near the anus,
from which it ejects the fetid liquor. The *Indians* value the fkins
highly, and ufe them as coverlets for their beds.

265. ZORRINA.   *Annas* of the Indians, *Zorrinas* of the | Mariputa, Mafutiliqui. *Gumilla Ore-*
Spaniards. *Garcilaffo de la Vega,* | *noque,* iii. 240. *De Buffon, Schreber,*
331. | cxxiii.

W with the back and fides marked with fhort ftripes of
  • black and white; the laft tinged with yellow: tail
long and bufhy; part white, part black: legs and belly black.
Lefs than the preceding.

Inhabits *Peru,* and other parts of *S. America:* its peftilential
vapour overcomes even the panther of *America,* and ftupefies that
formidable enemy.

266. RATEL.   Viverra Ratel. *Sparman Stock. Wettfk.* | Stink-bingfem. *Kolben,* ii. 133.
Hondl. 1777, 148. tab. iv. | Blaireau puant. *Voy. de la Caille,* 182.

W with a blunt black nofe: no external ears; in their place,
  • only a fmall rim round the orifice: tongue rough: legs
fhort: claws very long: ftrait, like thofe of a badger, and guttered beneath: color of the forehead, crown, and whole upper
part of the body, of a cinereous grey: cheeks, and fpace round
                                                              the

the ears, throat, breaft, belly, and limbs, black: from each ear
to the tail extends along the fides a dufky line, leaving beneath
another of grey.

Length from nofe to tail forty inches: of the tail, twelve: SIZE.
fore claws, an inch and three quarters long: hind claws one
inch.

Inhabits the Cape of *Good Hope*; lives on honey, and is a great   PLACE.
enemy to bees, which in that country ufually inhabit the deferted
burrows of the *Æthiopian* boar, the porcupine, jackals, and
other animals: preys in the evening: afcends to the higheft parts   MANNERS.
of the deferts to look about, and will then put one foot be-
fore its eyes, to prevent the dazzling of the fun. The reafon of
its going to an eminence, is for the fake of feeing or hearing
the *honey-guide cuckoo**, which lives on bees, and, as it were,
conducts it to their haunts: the *Hottentots* profit of the fame
guide. This animal cannot climb; but when he finds the bees
lodged in trees, through rage at the difappointment, will bite
the bark from the bottoms: by this fign alfo, the *Hottentots* know
that there is a neft of bees above.

The hair is very ftiff, and the hide fo tough, probably formed
fo by nature, as a defence againft the fting of bees, that it is not
eafily killed. It makes a ftout refiftance by biting and fcratch-
ing, and the dogs cannot faften on its fkin. A pack which could
tear a middle-fized lion to pieces, can make no impreffion on the

---

* A new fpecies, very fond of honey, which by its noife directs men, as well as
this beaft, to the bees neft. *Sparman*, in Phil. Tranf. lxvii. 43.

K 2   hide

hide of this beaſt : by worrying, they will leave it for dead, yet without inflicting on it any wounds.

This ſeems to be the *Stink-bingſem* of *Kolben*, and *Blaireau-puant* of *La Caille*, which they brand for the horrible ſtench which it emits from behind, by breaking wind; but the *Abbé* ſays, it quickly diſcharges the noiſome air. Mr. *Sparman* is ſilent in reſpect to this circumſtance. The *Hottentots* call it *Ratel*.

267. MARIPUTO.   Viverra Mariputo. *Gm. Lin.* 88.

W. of a black color, with a white bed, reaching from the forehead to the middle of the back : no ears : length twenty inches ; tail nine.

Obſerved by *Mutis* in *New Spain*, about the mines of *Pampluna :* ſleeps in the day : forms deep boroughs : wanders about in the night : feeds on worms and inſects : is very ſwift.

268. CEYLON.   *Gm. Lin.* 89.

W. above grey, mixed with duſky : below white. Size of the martin.

Inhabits the *Philippine* iſles and *Ceylon*.

Gm. Lin. 90.

W with three dufky lines along the back: tail longer than
  the body, with the tip black.
Inhabits *Barbary*. Defcribed by PALLAS.

Cook's firft voy. iii. 626. Martin-cat. *Stockdale's Bot. Bay,* 176.

W with fhort rounded ears: color black; marked with oblong
  fpots on the body, neck, and tail; belly of a pure white:
length from the tip of the nofe to the bafe of the tail, eighteen
inches: tail tapers elegantly to a point, and is about the fame length
as the body.
Inhabits the *Weftern* fide of *New Holland*.

White's Bot. Bay, 181.

W with long ears erect: color brown; lighteft on the tail: tail
  about the length of the body, covered with long hairs, and
ending in a point: fize of a rat.
Inhabits *New Holland*. According to Mr. *White's* defcription the
teeth are fo anomalous as to render it difficult to reduce this animal
to any genus.

White's

272. Spotted
Tafa.

White's Bot. Bay, 185.

THIS, according to Mr. *White*'s account and figure, differs from the former only in having the body and fides marked with irregular white fpots: tail plain.

273. Musky.

W with nofe, lower part of the cheeks, legs, and end of the tail, black: on the middle of the cheeks is a white fpot: body cinereous, dafhed with yellow: fome obfcure dufky lines and fpots mark the body and lower part of the tail.

Inhabits *Bengal:* fmells of mufk.　Sir *Elijah Impey.*

274. Civet.

La Civette qu'on nommóit anciennement Hyæna. *Belon obf.* 94.
Zibettus. *Caii opufc.* 43.
Felis Zibethus. *Gefner quad.* 837.
Animal Zibethicum, mafc. et fœm. *Hernandez Mex.* 580, 581.
Civet Cat. *Raii fyn. quad.* 178.
Coati Civetta vulgo. *Klein quad.* 73.

Meles fafciis et maculis albis nigris et rufefcentibus variegata. *Briffon quad.* 186.
Viverra Zibetha. V. cauda annulata, dorfo cinereo nigroque undatim ftriato. *Lin. fyft.* 65.
La Civette. *De Buffon,* ix. 299. *tab.* xxxiv. *Schreber,* cxi. Lev. Mus.

W with fhort rounded ears: fky-blue eyes: fharp nofe; the tip black: fides of the face, chin, breaft, legs, and feet black; the reft of the face, and part of the fides of the neck, white, tinged with yellow: from each ear are three black

4　　　　　　　　　　　　　　　　　　　ftripes,

ftripes, ending at the throat and fhoulders: the back and fides cinereous, tinged with yellow, marked with large dufky fpots difpofed in rows: the hair coarfe; that on the top of the body longeft, ftanding up like a mane: the tail fometimes wholly black; fometimes fpotted near the bafe: length, from nofe to tail, about two feet three inches; the tail fourteen inches: the body pretty thick.

Inhabits *India** *, the *Philippine* ifles †, *Guinea* ‡, *Æthiopia* ||, and *Madagafcar* § : the famous drug *mufk*, or civet, is produced from an aperture between the privities and the anus, in both fexes, fecreted from certain glands. The perfons who keep them, procure the mufk by fcraping the infide of this bag twice a week with an iron *fpatula*, and get about a dram each time; but it is feldom fold pure, being generally mixed with fuet or oil, to make it more weighty: the males yield the moft; efpecially when they are previoufly irritated. They are fed, when young, with pap made of millet, with a little flefh or fifh; when old, with raw flefh: in a wild ftate prey on fowl.

PLACE.

---

* *Dellon's voy.* 82.　　† *Argenfola,* iii.　　‡ *Bofman,* 238, *Barbot.* 114.
|| *Rauwolf's Travels,* ii. 482.　　§ *Flacourt's Madagafcar,* 154; where it is called *Falanouc.*

ZIBET.

275. β Zibet.   Animal Zibethicum Americanum. *Her-*   Le Zibet. *De Buffon,* 299. *tab.* xxxi.
                nandez Mex. 538.   *Schreber,* cxii.
                Felis Zibethus. *Gefner quad.* 836.

W with short rounded ears: sharp long nose: pale cinere-
• ous face: head, and lower part of the neck, mixed
with dirty white, brown, and black; sides of the neck marked
with stripes of black, beginning near the ears, and ending at the
breast and shoulders: from the middle of the neck, along the
ridge of the back, extends a black line, reaching some way up the
tail: on each side are two others: the sides spotted with ash-
color and black: the tail barred with black and white; the black
bars broader on the upper side than the lower.

A variety first distinguished from the other by M. *de Buffon*;
but figured long before by *Hernandez* and *Gefner*: unknown in
Mexico*, till introduced there from the *Philippine* isles. These
animals seem not to be known to the antients.

276. Musk.   W with the upper part of the body cinereous, dashed with
              • yellow, and marked with some obscure dusky lines:
nose, part of the cheeks, legs, and end of the tail, black; on the
middle of the cheeks is a white spot.

PLACE.   Inhabits *Bengal:* has a very strong musky scent: described from
a drawing in Sir *Elijah Impey*'s collection.

                          * *Hernandez Nov. Hifp.* ii.
                                                                    La

Sonnerat, voy. ii. 144. tab. xci.

W. with a long nofe: fhort erect ears: the ground-color of the whole animal perlaceous grey: face black: above each eye four black fpots: from the hind part of the head are three black lines; one paffes down the hind part of the neck and one down each fide of the neck, and over part of the fhoulders: from the breaft another extends along the middle of the belly; three others begin at the fmall of the back, and reach to the tail: on the body and thighs are forty-one round black fpots: the tail annulated with black and grey: legs and feet black: fize of a common cat.

This animal lives by the chace: leaps with great agility from tree to tree: is very fierce: emits a ftrong mufky fmell, produced from a liquor which exudes from an orifice above the parts of generation. The *Malayes* collect it, and pretend that it ftrengthens the ftomach, and excites to love. The *Chinefe* efteem it highly on account of the laft quality; and buy it from the *Malayes.* Inhabits the peninfula of *Malacca.*

MANNERS.

278. GENET.

La Genette. *Belon obf.* 74.
Genetha. *Gefner quad.* 549, 550.
Genetta vel Ginetta. *Raii fyn. quad.* 201.
Coati, ginetta Hifpanis. *Klein quad.* 73.
Muftela cauda ex annulis alternatim al-
bidis et nigris variegata. *Briffon quad.*

186.
Viverra Genetta. V. cauda annulata, corpore fulvo-nigricante maculato. *Lin.*
fyft. 65.
La Genette. *De Buffon*, ix. 343. *tab.*
xxxvi. *Schreber,* cxiii. LEV. MUS.

W with ears a little pointed: flender body: very long tail: color of the body a pale tawny, fpotted with black; and the ridge of the back marked with a black line: the tail annulated with black and tawny: feet black: fometimes the ground color of the hair inclines to grey: about the fize of a martin; but the fur is fhorter.

PLACE.

Inhabits *Turky, Syria,* and *Spain*; frequents the banks of rivers; fmells of mufk, and, like the civet, has an orifice beneath the tail: is kept tame in the houfes at *Conftantinople,* and is as ufeful as a cat in catching mice.

279. PILOSELLO.

La Genette *de la France, de Buffon,* Suppl. iii. tab. xlvii. p. 236.

W with nofe of a deep brown: face and chin cinereous: a dark line up the forehead: under fide of the neck cinereous, mixed with ruft: back and whole body of the fame color, varied with irregular black fpots: outfide of the hind legs and thighs dufky: foles of the feet and upper part down to the claws, cloathed with down: tail tawny, annulated with black. Leffer than the common ferret.

Inhabits

Fossane —— N. 280.

Inhabits the rock of *Gibraltar*, and the mountains of *Ronda*: called by the *Spaniards Pilofello*; found alfo in *France*. After the famous victory near *Tours*, gained over the *Saracens* in 726 by *Charles Martel*, fuch quantities of rich garments, made of the fkins of thefe animals, were found, as to give occafion to the hero to eftablifh an order of knighthood called *L'Ordre de la Genette*. On the firft inftitution there were fixteen knights; among them were the moft illuftrious princes of the time. *Martel* himfelf was the fovereign. The collar confifted of the chains of gold, mixed with enamelled rofes of red; pendent was a genet of gold, enamelled with black and red. The order continued during the fecond race of kings. It is faid to have given way afterwards to the *Order of the Star*.

La Foffane. *De Buffon*, xiii. 163. *tab. xx. Schreber*, cxiv. Lev. Mus.   280. FOSSANE.

W • with a flender body: rounded ears: black eyes: body and legs covered with cinereous hair, mixed with tawny: from the hind part of the head, towards the back and fhoulders, extend four black lines: the whole under fide of the body of a dirty white: tail femi-annulated.

Inhabits *Madagafcar*, and *Guinea, Bengal, Cochin-china*, and the PLACE. *Philippine* ifles: is fierce, and hard to be tamed: in *Guinea* is called *Berbe*; by the *Europeans*, Wine-bibber, being very greedy of *Palm-wine* *: deftroys poultry: is, when young, reckoned very good to eat †.

* *Bofman*, 239.
† *Flacourt hift. Madagafcar*, 512; where it is called *Foffa*.

L 2   The

The fpecimen in the *Leverian Mufeum* differed in fo many re-fpects, that it is neceffary to give a full defcription of it.

W. with a white fpot on each fide of the nofe, and another beneath each eye: the reft of the nofe, cheeks, and throat, black: ears very large, upright, rounded, thin, naked, and black: fore-head, fides, thighs, rump, and upper part of the legs, cinereous: on the back are many long black hairs: on the fhoulders, fides, and rump are difperfed fome black fpots: tail black towards the end; near the bafe mixed with tawny, and flightly annulated with black: feet black: claws white.

Size of the Genet, to which it bears a great refemblance: tail of the length of the body.

Six cutting teeth, two canine, in each jaw.

Five toes on each foot; each toe connected by a ftrong web.

281. GREATER.

Lutra. *Agricolæ An. Subter.* 482. *Gefner quad.* 687. *Raii fyn. quad.* 187.
Wydra. *Rzaczinfki Polon.* 221.
Otter. *Klein quad.* 91.
Muftela Lutra. M. plantis palmatis nudis, cauda corpore dimidio breviore. *Lin.*

fyft. 66. Utter. *Faun. fuec.* No. 12.
Lutra caftanei coloris. *Briffon quad.* 201.
Le Loutre. *Belon Aquat.* 26. *De Buffon,* vii. 134. *tab.* xi. *Schreber,* cxxvi. A. B.
Otter. *Br. Zool.* i. N° 19. *Br. Zool. illuftr. tab.* c. LEV. MUS.

O. with fhort ears: eyes placed near the nofe: lips thick: whifkers large: the color a deep brown, except two fmall fpots each fide the nofe, and another beneath the chin: the throat and breaft cinereous: legs fhort and thick, and loofely joined to the body; capable of being brought on a line with the body, and performing the part of fins; each toe connected to the other by a broad ftrong web.

The ufual length, from the tip of the nofe to the bafe of the tail, is twenty-three inches; of the tail fixteen: the weight of the male otter, from eighteen to twenty-fix pounds; of the female, from thirteen to twenty-two. Mr. *Ives* fays that the otters of the *Euphrates* are no larger than the common cat.

SIZE.

Inhabits all parts of *Europe,* N. and N. E. of *Afia,* even as far as *Kamtfchatka;* is found in none of the *Aleutian* or *Fox Iflands,* except in the eafternmoft, which are fuppofed to be near to the new world: is found in *Chili*:* abounds in *North America,* particularly

PLACE.

* *Molina,* 253.

5

in

in *Canada*, where the moſt valuable furs of this kind are produced: dwells in the banks of rivers; burrows, forming the entrance of its hole beneath the water; works upwards towards the ſurface of the earth, and makes a ſmall orifice, or air-hole, in the midſt of ſome buſh: is a cleanly animal, and depoſits its excrements in only one place: ſwims and dives with great eaſe: very deſtructive to fiſh; if they fail, makes excurſions on land, and preys on lambs and poultry. Sometimes breeds in finks and drains; brings four or five young at a time: hunts its prey againſt the ſtream: frequents not only freſh waters, but ſometimes preys in the ſea; but not remote from ſhore: will give a ſort of loud whiſtle by way of ſignal to one another *: is a fierce animal; its bite hard and dangerous: is capable of being tamed, to follow its maſter like a dog, and even to fiſh for him, and return with its prey.

The *Latax* of *Ariſtotle*†; poſſibly a large variety of Otter ‡.

Siya

---

* *Leonard Baldner*, iii. 139. fig. This was the perſon whom Mr. *Willughby* calls a fiſherman on the *Rhine*, of whom, on his travels in 1663, he bought a moſt beautiful and accurate collection of drawings of birds, fiſh, and a few beaſts, frequenting that great river about *Straſbourg*, of which city *Leonard* ſtiles himſelf, fiſherman and burgher. The work is dated in 1653. If I may judge from the elegance of his dreſs, in the portrait prefixed to the firſt volume, it ſhould appear that he was a perſon of conſiderable wealth. A *German* MS. deſcription is placed oppoſite to each drawing. This valuable work is now in the poſſeſſion of EDWARD KING, *Eſq*; and had been bought by a relation of his out of the collection of Dr. MEAD.

† *Hiſt. An. lib.* viii. *c.* 5. vide *Br. Zool.* i. 86. 4to.

‡ Sir *Joſeph Banks*, on his return from *Newfoundland*, was ſo obliging as to communicate to me the following account of ſome animals ſeen by a gentleman who

went

Siya & Cariguibeiu. *Marcgrave Brafil.* Lutra Brafilienfis. *Raii fyn. quad.* 189.   282. BRASILIAN.
234. *Des Marchais,* iii. 306.   *Briffon quad.* 202.

O. with a round head like that of a cat: teeth feline: eyes fmall, round, and black: large whifkers: ears round: feet in form of thofe of a monkey, with five toes; the inner the fhorteft: claws fharp: tail reaching no lower than the feet; flat and naked *.

Hair foft, and not long; entirely black, excepting the head, which is dufky; and the throat, which is yellow.

Bulk of a middling dog. If the fame with the otters of *Gui-*   SIZE. *ana,* mentioned by M. *de Buffon,* it weighs from forty to a hundred pounds †.

Inhabits *Brafil, Guiana,* and the borders of the *Oronoko,* pro-   PLACE.

went on that voyage; which I take the liberty of inferting here, as they bear fome relation to the Otter in their way of life. He obferved, fitting on a rock, near the mouth of a river, five animals, fhaped like *Italian* grehounds, bigger than a fox, of a fhining black color, with long legs, and long taper tail. They often leaped into the water and brought up trouts, which they gave to their young which were fitting with them. On his appearing, they all took to the water and fwam a little way from fhore, kept their heads out of the water, and looked at him. An old Furrier faid, that he remembered the fkin of one fold for five guineas; and that the *French* often fee them in *Hare-Bay.*

* *Barrere Fr. Æquin.* 155.

† *Suppl.* iii. 158, 159.

vided

vided the *Guachi* of *Gumilla* be the fame [*]. *Marcgrave* fays that it is an amphibious animal; lives on fifh, and cruftaceous animals, fuch as cray-fifh; and is very dextrous in robbing the nets and wheels of what it finds in them: makes a noife like a young puppy. The flefh is reckoned delicate eating, and does not tafte fifhy, notwith-ftanding its food.

If this is the *Guachi,* as probably it is, it burrows on the banks of rivers, and lives in fociety: are extremely cleanly, and carry to a diftance the bones and reliques of the fifh they have been eat-ing. They go in troops; are very fierce, and make a ftrong de-fence againft the dogs; but if taken young are foon tamed.

283. LESSER.

Noerza. *Agricola An. Subter.* 485. *Gefner quad.* 768.
Latax. *Germ.* Nurtz. *nobis* Nurek. *Rzac-zinfki Polon.* 218.
Muftela Lutreola. M. plantis palmatis, hirfutis ore albo. *Lin. fyft.* 66. *Fennis,* Tichurt; *Suecis,* Mænk. *Faun. fuec.* Nº 13.
Norka. *Ritchkoff orenb. Topogr.* i. 295. *Schreber,* cxxvi.

O, with roundifh ears: white chin: top of the head hoary; in fome tawny: body tawny and dufky; the fhort hairs being yellowifh; the long hairs black: the feet broad, webbed,

[*] *Hift. de l'Orenoque,* iii. 239. *Gumilla* calls them alfo *Loups ou Chiens d'Eau,* and fays they are as large as a fetting-dog. There is a great difagreement in the form of the feet, with others of the Otter kind. The writers who have had opportunity of examining it, are filent about the webs, the character of the genus. Till that point is fettled, I muft remain doubtful whether it be the *Sa-ricovienne* of *Andrew Thevet,* as M. *de Buffon* conjectures. The fize of the lat-ter is another objection, which will apologize for my making a feparate article of that animal till this point is fettled.

and

P.Mazell sculp.

Lesser Otter — N.º 283.

and covered with hair: tail dufky, and ends in a point: of the form of an otter, but thrice as fmall.

Inhabits *Poland*, and the north of *Europe*; and is found on the banks of all the rivers in the country north of the *Yaik*. None are found beyond the lake *Baikal*, or in the north-eaft parts of *Siberia*. Lives on fifh, frogs, and water-infects: its fur very valuable; next in beauty to that of the fable. Caught in *Bafhkiria* with dogs and traps: is moft exceffively fœtid.

The *Minx* of *North America* is the fame animal with this. The late worthy Mr. *Peter Collinfon* * favored me with the following account he received from Mr. *John Bartram*, of *Penfylvania:*
' The *Minx*,' (fays he) ' frequents the water like the Otter, and
' very much refembles it in fhape and color, but is lefs; will
' abide longer under water than the mufk quafh, mufk rat, or
' little beaver: yet it will leave its watery haunts to come and
' rob our hen-roofts; bites off their heads and fucks their blood:
' when vexed, it has a ftrong loathfome fmell; fo may be called
' the *Water Pole Cat:* its length, from nofe to tail, twenty inches;
' the tail four: is of a fine fhining dark brown color.'

From the conformity between the names this animal goes by, in *America* and *Sweden* (*Minx* and *Mænk*) it feems as if fome

* By letter dated *June* 14, 1764. *Lawfon* alfo gives fome account of it, p. 122, *Hift. Carolina:* He fays it is a great enemy to the Tortoifes; whofe eggs it fcrapes out of the fand and devours: eats frefh-water mufcles, whofe fhells are found in great abundance at the mouth of their holes, high up in the rivers, in whofe banks they live: may be made domeftic: is a great deftroyer of rats and mice. *La Hontan*, i. 232, feems to mean the fame animal, by his *Foutereaux*, an amphibious fort of little Pole-cats.

Swedish colonift, who had feen it in his own country, firft beftowed the name it now goes by, a little changed from the original: the fkins are often brought over to *England*.

2847 CHINCHI-MEN.   *Molina Chili, 265.*

O. with head, whifkers, ears, eyes, fhape, and length of the tail, exactly refembling the domeftic cat: feet furnifhed with five toes, palmated, and with ftrong and crooked claws: body covered with two forts of hair, one very fhort and fine, the other long and rude: length from nofe to tail twenty inches.

MANNERS.   Inhabits the fea of *Chili*, and very feldom quits that element: goes always in pairs: loves to bafk in the fun: creeps to the fummit of the rocks, where it is taken in traps: has a hoarfe voice, and all the fiercenefs of the wild cat.

85. SARICO-VIENNE.

O. of the fize of a cat, with a fur fine as velvet, grey and black: web footed.

Lives more in the water than on land: the flefh very delicate, and good to eat.

This appears to me to be the very fame with *La petite Loutre d'eau douce de Cayenne,* defcribed and figured by M. *de Buffon*[*], probably from a young animal.

[*] *Suppl.* iii. 159. tab. xxii.

The

The body, fays he, is feven inches (*French*) in length : the tail fix inches and feven lines; flender, taper, tuberculated, convex above, flat beneath : ears rounded, and longer than ufual with otters : head, cheeks, and back, dufky; and the fides marked regularly with the fame colors, iffuing from the back, extending almoft to the belly; the fpaces between of a yellowifh grey : above each eye is a white fpot : the throat, and whole under fide of the body, of the fame color : the toes before are divided; thofe behind webbed.

M. *de la Borde*, as quoted by M. *de Buffon*, mentions another fpecies of Otter frequent in the rivers of *Guiana*, weighing from twenty to twenty-five pounds, and of a yellowifh color.

Muftela Lutris. M. plantis palmatis pilofis, cauda corpore quadruplo breviore *Lin. fyft.* 66. *Schreber*, cxxviii.
Lutra marina, Kalan. *Nov. Com. Petrop.*

ii. 367. *tab.* xvi.
Sea Otter. *Hift. Kamtfchatka*, 122. *Muller's voy.* 57, 58.

O. with a black nofe : upper jaw longer and broader than the lower : long white whifkers : irides hazel : ears fmall, erect, conic : in the upper jaw are fix cutting teeth; in the lower four : the grinders broad, adapted for breaking and comminuting cruftaceous animals, and fhell-fifh : fkin thick : hair thick and long, exceffively black and gloffy : beneath that a foft down : color fometimes varies to filvery : legs thick and fhort : toes covered with hair, and joined by a web : the hind feet exactly like thofe of a feal, and have a membrane fkirting the out-

M 2

fide

SIZE.

fide of the exterior toe, like that of a goofe　Length from nofe to tail is ufually above three feet; but there have been inftances of fome being a foot longer: the tail thirteen inches and a half; flat, fulleft of hair in the middle; fharp-pointed.　The biggeft of thefe animals weigh feventy or eighty pounds.

PLACE.

Inhabits, in vaft abundance, *Bering's* ifland, *Kamtfchatka,* the *Aleutian* and the *Fox iflands* between *Afia* and *America,* and in the interior fea as far as has been difcovered to the eaft of *De Fuca's* ftreights.　They are fometimes feen in troops of hundreds, and a hundred leagues from land.　They are entirely confined between lat. 49. and 60 north; and between eaft long. from *London* 126 to 150. During winter they are brought in great numbers by the eaftern winds from the *American* to the *Kurilian* iflands.

MANNERS.

Are moft harmlefs animals: moft affectionate to their young; will pine to death at the lofs of them, and die on the very fpot where they have been taken from them: before the young can fwim, they carry them in their paws, lying in the water on their backs: run very fwiftly; fwim often on their back, their fides, and even in a perpendicular pofture: are very fportive; embrace each other, and even kifs: inhabit the fhallows, or fuch which abound with fea-weeds: feed on lobfters, fifh, *Sepia,* and fhell-fifh: breed once a year; bring but one young at a time; fuckle it a year, bring it on fhore: are dull fighted, but quick fcented: hunted for their fkins, which are of great value; fold to the *Chinefe* for feventy or a hundred *rubles* apiece: each fkin weighs three pounds and a half.　The young are reckoned very delicate meat, fcarcely to be diftinguifhed from a fucking lamb.

LENGTH

LENGTH from nofe to tip of tail four feet four inches : of the tail about thirteen inches : diameter of body fcarcely more than five inches and a half : fore legs about three inches and a half long : hind legs about four inches : head fmall, eyes fmall, ears moft extremely fmall, fcarce vifible : fore feet webbed; hind feet more ftrongly fo : color of the whole animal a rich very deep chefnut or dark brown, rather paler beneath : cheeks and throat paler than the other parts, or more inclining to whitifh.

Inhabits *Staten-Land.*

287. SLENDER.

PLACE.

DIV.

DIV. II. Sect. III,

DIGITATED QUADRUPEDS.

Without canine teeth; and with two cutting teeth in
  each jaw.
enerally herbivorous, or frugivorous.

DIV. II. Sect. III. Digitated Quadrupeds.

XXV. CAVY. Two cutting teeth in each jaw.

Generally four toes on the fore feet, three behind.

Short ears: no tail, or a very fhort one.

Pace creeping; and flow: numerous breeders: fhort-lived.

288. Capibara.

Caby-bara. *Marcgrave Brafil.* 230. *Pifo Brafil.* 99. *Raii fyn. quad.* 126.
River hog. *Wafer in Dampier*, iii. 400.
Cochon d'Eau. *Des Marchais*, iii. 314.
Susmaximuspaluftris. Cabiai, cabionora. *Barrere France Æquin.* 160.

Capivard. *Froger's voy.* 99.
Sus hydrochæris. S. plantis tridactylis cauda nulla. *Lin. fyft.* 103.
Hydrochærus, Le Cabiai. *Briffon quad.* 80. *De Buffon*, xii. 384. *tab.* xlix.
Irabubos. *Gumilla Orenoque*, iii. 238.

C. with a very large and thick head and nofe: fmall rounded ears: large black eyes: upper jaw longer than the lower: two ftrong and great cutting teeth in each jaw: eight grinders in each jaw; and each of thofe grinders form on their furface feemingly three teeth, each flat at their ends *: legs fhort: toes long, connected near their bottoms by a fmall web; their ends guarded by a fmall hoof: no tail: hair on the body fhort, rough, and

* M. *de Buffon* denies this: his defcription was taken from a young fubject; but *Marcgrave* and *Des Marchais*, who had opportunities of examining thefe animals in their native country, agree in this fingular conftruction of the teeth.

brown

brown; on the nofe, long and hard whifkers: grows to the fize of a hog of two years old.

Inhabits the country from the Ifthmus of *Darien* to the *Brafils*, and even to *Paraguay*; lives in the fenny parts, not remote from the banks of great rivers, fuch as the *Oronoque, Amazons,* and *Rio de la Plata:* runs flowly: fwims and dives remarkably well, and keeps for a long time under water: feeds on fruits and vege-tables: is very dextrous in catching fifh, which it brings on fhore, and eats at its eafe: it fits up, and holds its prey with its fore feet, feeding like an ape: feeds in the night, and commits great ravages in gardens: keeps in large herds, and makes an horrible noife like the braying of an afs: grows very fat: the flefh is eaten, is tender, but has an oily and fifhy tafte: is eafily made tame *, and foon grows very familiar.

Cuniculus vel Porcellus Indicus. *Gefner quad.* 367.
Cavia Cobaya. *Marcgrave Brafil.* 224. *Pifo Brafil.* 102.
Mus feu cuniculus *Americanus et Guineen-fis,* Porcelli pilis et voce, Cavia Cobaya. *Raii fyn. quad.* 223.
Cavia Cobaya *Brafil.* quibufdam mus Pharaonis. Tatu pilofus. *Klein quad.*

49.
Mus porcellus. M. cauda nulla, palmis tetradactylis, plantis tridactylis. *Lin. fyft.* 79. *Amœn. Acad.* iv. 190. *tab.* ii.
Cuniculus ecaudatus, auritus albus, aut rufus, aut ex utroque variegatus. *Briffon quad.* 102.
Le Cochon d'Inde. *De Buffon,* viii. 1. *tab.* i. LEV. MUS.

289. RESTLESS.

C. with the upper lip half divided: ears very large, broad, and rounded at the fides: hair erect, not unlike that of a young pig; color white, or white varied with orange and black, in irre-gular blotches: no tail: four toes on the fore feet; three on the hind.

* *Muratori hift. Paraguay,* 258.

C A V Y.

Inhabits *Brafil:* no mention made by writers of its manners in a wild ftate: domefticated in *Europe:* a reftlefs, grunting, little animal; perpetually running from corner to corner: feeds on bread, grains, and vegetables: breeds when two months old: brings from four to twelve at a time; and breeds every two months: would be innumerable, but numbers of the young are eaten by cats, others killed by the males: are very tender, multitudes of young and old perifhing with cold: are called in *England, Guinea Pigs,* being fuppofed to come from that country. Rats are faid to avoid their haunts.

290. ROCK.

Aperea. *Brafilienfibus,* nobis Veldratte, vel Bofchratte. *Marcgrave Brafil.* 223. *Pifo Brafil.* 103. *Raii fyn. quad.* 206. Cavia Aperea. *Klein quad.* 50.

Cuniculus ecaudatus auritus, ex cinereo rufus. *Briffon quad.* 103. L'Aperea. *De Buffon,* xv. 160. LEV. Mus.

C. with divided upper lip: fhort ears: four toes on the fore feet; three on the hind: no tail: color of the upper part of the body black, mottled with tawny: throat and belly white: length one foot.

Inhabits *Brafil:* lives in the holes of rocks: is driven out, and taken by little dogs: is fuperior in goodnefs to our rabbets: its paces like thofe of a hare.

Narborough's

Patagonian Cavy —— *N.º 291.*

Narborough's voy. 33. LEV. MUS.

C. with long ears, much dilated near the bottom: upper lip divided: on each fide of the nofe tufts of foft hairs, and long whifkers: tip of the nofe black: face, back, and fore.part of the legs, cinereous and ruft-colored: breaft and fides tawny: belly of a dirty white: on each thigh a white patch: rump black: legs very long: claws long, ftrait, and black; four on the fore feet; three on the hind: tail a mere naked ftump.

This animal is found of the weight of fix-and twenty pounds *.

Is found in plenty about *Port Defire*, in *Patagonia*: lives in holes of the earth, like the rabbet: the flefh of a fnowy whitenefs, and excellent flavor †.

Sir *John Narborough*, and other voyagers, call it a hare.

Paca. *Marcgrave Brafil.* 224. *Pifo Brafil.* 101. *De Laet*, 484.
Mus Brafilienfis magnus, porcelli pilis et voce, Paca dictus. *Raii fyn. quad.* 226.
Cavia Paca. *Klein quad.* 50.
Cuniculus major, paluftris, fafciis albis notatus. Paca *Marcgrave. Barrere France Æquin.* 152.
Mus Paca. M. cauda abbreviata, pedibus pentadactylis, lateribus flavefcenti-lineatis. *Lin. fyft.* 81.
Cuniculus caudatus, auritus, pilis obfcurè fulvis, rigidis, lineis ex albo flavefcentibus ad latera diftinctis. *Briffon quad.* 99.
Le Paca. *De Buffon*, x. 269. *tab.* xliii. *Supplem.* iii. 203. *tab.* xliii. LEV. MUS.

C. with the upper jaw longer than the lower: noftrils large: whifkers long: ears fhort and naked: neck thick: hairs fhort and hard: color of the upper part of the body dark

---

* *Byron's voy.* 18.　　　† The fame, 19.

N 2

brown;

brown; the lower part, or fides, marked lengthways with lines of grey fpots: the belly white; in fome, perhaps young ones, the fides and fpots are of a pale yellow: five toes on each foot: only the meer rudiment of a tail: length about ten inches: is made like a pig, and in fome parts is called the *Hog-Rabbet**.

Inhabits *Brafil*, and *Guiana*: lives in fenny places: burrows under ground: grows very fat: is efteemed in *Brafil* a great delicacy: grunts like a pig: eats its meat on the ground, not fitting up, as fome others of this genus do: are difcovered by little dogs, who point out the places they lie in: the mafter digs over them, and when he comes near transfixes them with a knife; otherwife they are apt to efcape: will bite dreadfully. There is a variety quite white, found on the banks of the river *St. Francis†*.

293. BRISTLY.

Agnus filiorum Ifrael. *Profp. Alp. Ægypt.* i 232.
Daman Ifrael. *De Buffon Suppl.* vi. 276. *tab.* xlii.

Afhnoko. *Bruce's travels*, v. 139.
Hirax Syriacus. *Gmel. Lin. fyft.* 167. *Schreber, tab.* ccxi. B.

C. with fhort oval ears, covered within and without with hair: color of the whole animal above grey and ferruginous: from the chin to the extremity of the belly white: on the upper a ftrong brifly mufticho, three inches five eighths long; above the eyes another tuft, two inches and two eighths long; all over the body are fcattered fimilar briftles, two inches and a quarter in length: the toes are flefhy; the lower part naked, the upper covered with

* *Wafer's voy. in Dampier*, iii. 401.   † *De Laet*, 484.

black

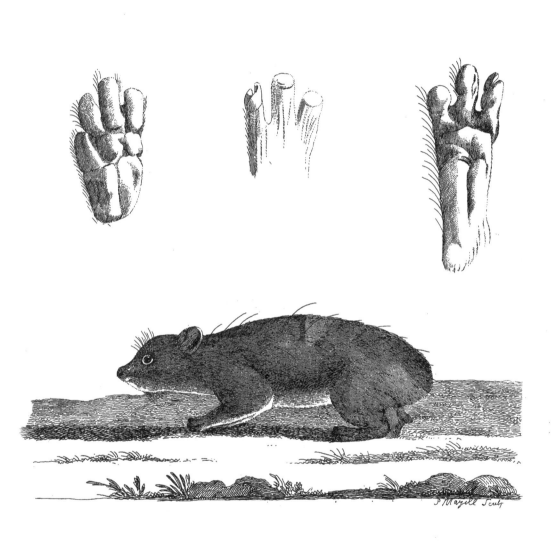

Bristly Cavy — Nº 293

P. Mazell Sculp

black hairs : the claws fomewhat refemble nails, and are ill adapted for burrowing: no tail: the length of the whole animal is
about feventeen inches.

This fpecies was firft taken notice of by *Profper Alpinus,* who calls
it *Agnus filiorum Ifrael*; the *Daman Ifrael* of the *Arabs.* He fays
it is larger than a rabbet, an object of the chace, and that the flefh
is fweeter than that of the rabbet.

PLACE.

Inhabits, according to Mr. *Bruce,* mount *Libanus,* the mountain
of the Sun in *Abyffinia,* and in great numbers Cape *Mahomet,* on
the *Arabian* gulph, not far to the eaft of *Suez.* By *Alpinus* we find
they are alfo inhabitants of *Ægypt.* They are gregarious, and fit
by dozens on the great ftones to bafk in the fun, before the mouth
of caves, or clefts in the rocks, their places of refuge at the fight
of man. They are juftly fuppofed by Mr. *Bruce* to have been
the *Saphen* (miftranflated the coney) of HOLY WRIT. *Solomon*
fays, ' The hills are the refuge for the wild goats, and the rocks for
the *Conies.* See his *Saphen.* ' The *Saphen,* adds he, are but a feeble
' folk, yet make they their houfes in the rocks*.' They retire
into the depths of the clefts, and there make themfelves a houfe;
i. e. a neft of ftraw. Neither the *Chriftians* of *Abyffinia* and the
Mahometans, eat the flefh of thefe animals. The *Arabs* of mount
Libanus and of *Arabia Petræa* ufe them as a food. The flefh is as
white as a chicken, and free from any ranknefs.

Mr. *Bruce* fuppofes that Doctor *Shaw* intended this animal by his
Jird†; but, as our learned countryman exprefsly fays that his

---

* *Proverbs,* ch. xxx. v. 24, 26.
† *Shaw's* Travels, p. 248.

4   animal

animal has a tail, and that only a little fhorter than that of the common rat, we muft have recourfe to fome other fpecies, perhaps genus, for the *Jird* of *Barbary*.

294. LONG-NOSE.

Aguti vel Acuti. *Marcgrave Brafil.* 224. *Pifo Brafil.* 102.
Acuti ou Agoutis. *De Laet*, 484. *Rochefort Antilles*, i. 287.
Mus fylveftris Americanus cuniculi magnitudine, pilis et voce Porcelli, Aguti. *Raii fyn. quad.* 226.
Cavia Aguti. M. cauda abbreviata, palmis tetradactylis, plantis tridactylis, abdomine flavefcente. *Lin. fyft.* 80.

Cuniculus caudatus, auribus, pilis ex rufo et fufco mixtis rigidis veftitus. *Briffon quad.* 98.
L'Agouti. *De Buffon*, viii. 375. *tab.* l.
Small Indian Coney. *Brown's Jamaica*, 484.
Long-nofed Rabbet. *Wafer's voy. in Dampier*, iii. 401.
Cuniculus omnium vulgatiffimus, Aguti vulgo. *Barrere France Æquin.* 153 *.

C· with a long nofe: divided upper lip: fhort rounded ears: black eyes: hard and fhining; on the body mixed with red, brown, and black; on the rump, of a bright orange-color: belly yellow: legs almoft naked, flender, and black: four toes on the fore feet; three on the hind; tail fhort, and naked: fize of a rabbet.

Inhabits *Brafil*, *Guiana*, &c. Grunts like a pig: is very voracious: fits on its hind legs, and holds its food with the fore feet when it eats: hides what it cannot confume: hops like a hare: goes very faft: when purfued, takes fhelter in hollow trees: is capable of being tamed: when angry, fets up the hair on its

* The animal defcribed by *Seba* under the name of *Cuniculus Americanus*, i. 67. *tab.* xli. feems the fame with this, notwithftanding he fays, that the hind feet are tetradactylous.

back,

back, and ftrikes the ground with its feet: is eaten by the inhabitants of *South America.*

<table>
<tr><td>Cuniculus minor caudatus, olivaceus, Akouchy. *Barrere France Æquin.* 153. *Des Marchais,* iii. 303.</td><td>L'Akouchy. *De Buffon,* xv.258. *Suppl.*iii. 211. *tab.* xxxvi.</td><td>295. OLIVE.</td></tr>
</table>

A Species of *Aguti,* lefs than the former, and of an olive-color: which is the whole account left us by M. *Barrere. Des Marchais* fays, it is more delicate food than the other.

Inhabits *Guiana,* and the iflands of St. *Lucia* and *Grenada :* inhabits the woods: lives on fruits: is excellent meat: its flefh is white: eafily made tame: makes a cry (but very rarely) like the *reftlefs cavy:* abhors water.

<table>
<tr><td>Java hare. *Catefby Carolina, App. tab.* xviii.
Cavia Javenfis. *Klein quad.* 50.
Cuniculus caudatus auritus, rufefco ad-</td><td>mixto. *Briffon quad.* 98.
Mus leporinus. *Lin. fyft.* 80.
Cuniculus Americanus. *Seb. Muf.* i. 67. tab. xlii. fig. 2.</td><td>296. JAVAN.</td></tr>
</table>

C. with a flender fmall head: prominent naked ears, rounded at the tops: hairs very ftiff like briftles, efpecially on the back: color of the upper part of the body reddifh: breaft and belly white: legs long: hind parts large: four toes on the fore feet; three on the hind: tail fhort: fize of a hare.

Inhabits *Surinam* and the hotter parts of *South America,* where it is a common food: the flefh is white, but dry. It is not
found

found in *Java* or *Sumatra*, as *Catefby* afferts. Governor *Loten* affures me, that he made the moft diligent enquiry after it in moft parts of *Java*, but could never find the left traces of any fuch animal.

297. CAPE.   Cavia capenfis. *Pallas Mifcel. Zool.* 30.   *Monogr. De Buffon, Supplem.* iii. 177.
      tab. ii. *Spicil.* 16. *tab.* ii.   tab. xxix.
      Africaanfch bafterd-mormeldier. *Vofmaer*

C. with a thick head, and full cheeks: ears oval, half hid in the fur: head of the color of a hare: along the top of the back dufky, mixed with grey; fides and belly of a whitifh grey: four toes on the fore feet, three behind: tail fcarce vifible: fize of a rabbet, but the fhape of the body thick and clumfy.

Inhabits in great abundance the rocky mountains near the *Cape of Good Hope*, where it is called *Kaapfche Dafs, Klip Dafs**, or *Cape Badger:* burrows under ground: has a flow creeping pace; a fharp voice, often repeated: is efteemed very good meat.

  * *Kolben, Dutch edition*, as quoted by Dr. *Pallas*. *La Caille* mentions this fpecies under the name of *Marmot*.

Les Rats mufquès, Piloris. *Rochefort An-*   302. *De Buffon*, x. 2.   298. Musk.
tilles, i. 288. *Du Tertre hift. Antilles*, ii.

C of a black or tan color on the upper part of its body: white on the belly : tail very fhort * : almoft as big as a rabbet.

Inhabits *Martinico* and the reft of the *Antilles :* burrows like a rabbet : fmells fo ftrong of mufk, that its retreat may be traced by the perfume : an obfcure fpecies, never examined by a na-turalift.

    * *Nouv. voy. aux ifles de l'Amerique*, i. 438.

XXVI. HARE.   Two cutting teeth in each jaw.

Short tail: or none.

Five toes before; four behind.

299. COMMON.   Lepus. *Plinii lib.* viii. *c.* 55. *Gefner quad.* 605. *Raii fyn. quad.* 204.

Hafe, *Klein quad.* 51.

Lepus timidus. L. cauda abbreviata auriculis apice nigris? *Lin. fyft.* 77. Hafe, *Faun. fuec.* No. 25.

Lepus caudatus ex cinereo rufus. *Briffon quad.* 94.

Le lievre. *De Buffon,* vi. 246. *tab.* xxxviii. *Br. Zool* i. N° 20.

Arnæb. *Forfkal.* iv. LEV. MUS. in which are feveral curious varieties of colored hares.

H. with ears tipt with black: eyes very large and prominent: chin white: long white whifkers: hair on the face, back, and fides, white at the bottom, black in the middle, and tipt with tawny red: throat and breaft red: belly white: tail black above, white beneath: feet covered with hair even at the bottom: a large hare weighs eight pounds and a half. I am informed, that in the *Ifle of Man* fome have been known to weigh twelve: its length, from the nofe to the tail, two feet.

Inhabits all parts of *Europe,* moft parts of *Afia, Japan, Ceylon* *, *Ægypt* †, and *Barbary* ‡: a watchful, timid animal: always lean: fwifter in running up hill than on even ground: when ftarted, immediately endeavours to run up hill: efcapes the hounds by various artful doubles: lies the whole day on its feat: feeds by night: returns to its form by the fame road that it had

* *Kampfer Japan,* i. 126. *Knox Ceylon,* 20.   † *Profp. Alp.* i. 232.

‡ *Shaw's Travels,* 249.

taken

taken in leaving it: does not pair: the rutting-feafon is in *Fe-bruary* or *March*, when the male purfues the female by the fa-gacity of its nofe: breeds often in the year; brings three or four at a time: are very fubject to fleas: the *Dalecarlians* make a cloth of the fur, which preferves the wearer from their attacks: the fur is of great ufe in the hat manufacture: feeds on vege-tables: fond of the bark of young trees: a great lover of birch, parfly and pinks: was a forbidden food among the *Britons:* the *Romans*, on the contrary, held it in great efteem.

> *Inter quadrupedes gloria prima lepus,*

was the opinion of *Martial*; and *Horace*, who was likewife a *Bon vivant,* fays, that every man of tafte muft prefer the wing

> *Fæcundi leporis fapiens fectabitur armos.*

There have been feveral inftances of what may be called mon-fters in this fpecies, *horned hares*, excrefences growing out of their heads, likeft to the horns of the roe-buck. Such are thofe figured in *Gefner*'s hiftory of quadrupeds, p. 634; in the *Mufeum Regium* HAFNIÆ, No. 48. tab. iv; and in *Klein*'s hiftory of qua-drupeds, 32. tab. iii; and again defcribed in *Wormius's Mufeum,* p. 321, and in *Grew's Mufeum* of the Royal Society. Thefe in-ftances have occurred in *Saxony*, and I think in *Denmark*, to which may be added another near *Aftracan**.

HORNED HARES.

A farther account of two ftraw-colored animals like dogs, which run like hares, and were of the fame fize, feen by the late navigators in *New Holland*†, will, I fear, be a long *defideratum* among naturalifts.

* *Pallas.* † *Cook's voy.* iii. 565.

300. VARYING.   Lepus hieme albus. *Forſter hiſt. nat.* VOL-   Lepus variabilis. *Pallas. nov.ſp.* i. LEV.
                 GÆ. *Ph. Tranſ.* lvii. 343.   MUS.
                 Alpine hare. *Br. Zool.* i. N° 20.

H. with ſoft hair, in ſummer grey, with a ſlight mixture of
black and tawny: with ſhorter ears, and more ſlender
legs, than the common hare: tail entirely white, even in ſummer: the feet moſt cloſely and warmly furred. In winter, the
whole animal changes to a ſnowy whiteneſs, except the tips and
edges of the ears, which remain black, as are the ſoles of the
feet, on which, in *Siberia*, the fur is doubly thick, and yellow.
Leſs than the common ſpecies.

PLACE.   Inhabits the higheſt *Scottiſh Alps*, *Norway*, *Lapland*, *Ruſſia*,
Siberia *, *Kamtſchatka*, and the banks of the *Wolga*, and *Hudſon's*
Bay. In *Scotland*, keeps on the tops of the higheſt hills; never
deſcends into the vales; never mixes with the common hare,
which is common in its neighborhood: does not run faſt: apt
to take ſhelter in clefts of rocks: is eaſily tamed: full of frolic:
fond of honey and carraway comfits: eats its own dung before
a ſtorm: changes its color in *September:* reſumes its grey coat
in *April:* in the extreme cold of *Greenland* only, is always †
white. Both kinds of hares are common in *Siberia*, on the
banks of the *Wolga*, and in the *Orenburg* government. The one
never changes color: the other, native of the ſame place, con-

* Vide *Pontop. Norway*, ii. 9. *Scheffer Lapland*, 137. *Strahlenberg Ruſſia*, 370.
Ritchkoff Orenberg Topog. i. 287.
† *Egede, Greenl.* 62. *Crantz Greenl.* i. 70.

ſtantly

1. Varying Hare __ N°.300 __ 11. Hooded Rabbet __ p.104.

ſtantly aſſumes the whiteneſs of the ſnow during winter. This it does, not only in the open air, and in a ſtate of liberty : but, as experiment has proved, even when kept tame, and preſerved in houſes in the ſtove-warmed apartments; in which it experiences the ſame changes of colors as if it had dwelt on the ſnowy plains *.

They collect together, and are ſeen in troops of five or ſix hundred, migrating in ſpring, and returning in autumn †. They are compelled to this by the want of ſubſiſtence, quitting in the winter the lofty hills, the ſouthern boundaries of *Siberia*, and ſeek the plains and northern wooded parts, where vegetables abound; and towards ſpring ſeek again the mountainous quarters ‡. Mr. *Muller* ſays, he once ſaw two black hares, in *Siberia*, of a wonderful fine gloſs, and of as full a black as jet. Near *Caſan* was taken another in the middle of the winter 1768. Theſe ſpecimens were much larger than the common kind.

In the ſouthern and weſtern provinces of *Ruſſia* is a mixed breed of hares, between this and the common ſpecies. It ſuſtains, during winter only, a partial loſs of color: the ſides, and more expoſed parts of the ears and legs, in that ſeaſon, become white; the other parts retain their colors. This variety is unknown beyond the *Urallian* chain. It is called by the *Ruſſians, Ruſſak*; they take them in great numbers in ſnares, and export their ſkins to *England* and other places, for the manufacture of hats ||. The *Ruſſians* and *Tartars*, like the *Britons* of old, hold

MIGRATIONS.

BLACK HARES.

α. SPURIOUS.

* *Pallas* nov. ſp. faſc. i. p. 7.   † *Bell's Travels*, i. 238.   ‡ *Pallas* nov. ſp. faſc. i. p. 15. || The ſame, p. 6.

the flesh of hares in detestation, esteeming it impure: that of the VARIABLE, in its white state, is excessively insipid.

Hare, hedge Coney. *Lawson*, 122. *Catesby*, App. xxviii.

H. with the ears tipt with grey: upper part of the tail black; lower white: neck and body mixed with cinereous, rust-color, and black: legs of a pale ferruginous: belly white; fore legs shorter, hind legs longer, in proportion, than those of the common hare.

Length eighteen inches: weighs from three to four pounds and a half.

Inhabits all parts of *North America*. In *New Jersey*, and the colonies south of that province, it retains its color the whole year. In *New England**, *Canada*, and about *Hudson's Bay*, at approach of winter, it changes its short summer's fur for one very long, silky, and silvery, even to the roots of the hairs; the edges of the ears only preserving their color: at that time it is in the highest season for the table†; and is of vast use to those who winter in *Hudson's Bay*, where they are taken in vast abundance, in springes made of brass wire, to which the animals are led by a hedge made for that purpose, with holes left before the snares for the rabbets to pass through.

They breed once or twice a year, and have from five to seven at a time: they do not migrate, like the preceding, but always haunt the same places: they do not burrow, but lodge under

---

* *Josslyn's* Rarities, 22.   † *Clerk Californ.* i. 156.

fallen timber, and in hollow trees: they breed in the grafs; but in fpring fhelter their young in the trees, to which they alfo run when purfued; from which, in the fouthern colonies, the hunters * force them by means of a hooked ftick, or by making a fire, and driving them out by the fmoke. I have had an opportunity of examining this fpecies in its brown drefs from *Penfyl-vania,* and its winter's drefs from *Hudfon's Bay.*

Cuniculus. *Plinii. lib.* viii. *c.* 55. *Gefner quad.* 362. *Agrcola An Subt.* 482 Rabbet, or Coney. *Raii fy. quad.* 205. Lepufculus, cuniculus terram fodiens, Kaninchen. *Klein quad.* 52. Lepus cuniculus. L. cauda abbreviata, auriculis nudatis. *Lin. fyft.* 77.

Kanin. *Faun. fuec.* No. 26. *Br. Zool.* i. N° 22. Lepus caudatus, obfcuré cinereus. *Briffon quad.* 95. Le Lapin *De Buffon,* vi. 303. *tab.* l. li. Lev. Mus.

302. Rabbet.

H. with ears almoft naked: color of the fur, in a wild ftate, brown; tail black above, white beneath: in a tame ftate, varies to black, pied, and quite white: the eyes of the laft of a fine red.

Inhabits, in a wild ftate, the temperate and hot parts of *Europe,* and the hotteft parts of *Afia* and *Africa:* not originally *Britifh;* but fucceeds here admirably: will not live in *Sweden,* or the northern countries, except in houfes. *Strabo* † tells us, that they were firft imported into *Italy* from *Spain.* Not natives of *America;* but encreafe greatly in *S. America.*

Moft prolific animals: breed feven times in a year: produce eight young at a time: fuppofing that to happen regularly, one

* *Kalm,* ii. 45.   † *Lib.* ii.

pair

pair may bring in four years 1,274,840. In warrens, keep in their holes in the middle of the day; come out morning and night: the males apt to deftroy the young: the fkins a great article of commerce; numbers exported to *China*: the fur of great ufe in the hat-manufacture.

β. ANGORA RABBET. With hair long, waved, and of a filky finenefs, like that of the goat of *Angora*, vol. i. p. 61, and the *Cat*, vol. i. p. 296.

γ. HOODED RABBET. With a double fkin over the back, into which it can withdraw its head: another under the throat, in which it can place its fore feet: has fmall holes in the loofe fkin on the back, to admit light to the eyes: color of the body cinereous: head and ears brown.

Defcribed from a drawing, and manufcript account, by Mr. *G. Edwards*, preferved in the *Mufeum*; infcribed " A *Ruffian Rab-* " *bet*;" but I find that it is unknown in that empire.

303. BAIKAL.   Cuniculus infigniter caudatus, coloris Le-   Lepus cauda in fupina parte nigra in prona
              porini. *Nov. Com. Petrop.* v. 357. *tab.*   alba. *Briffon quad.* 97.
              xi.   Le Tolai. *De Buffon,* xv. 138.

H. with a tail longer than that of a rabbet: ears longer in the male, in proportion, than thofe of the *varying* hare: fur of the color of the common hare: red about the neck and feet:

tail

tail black above, white beneath : fize between that of the common and the *varying* hare.

Inhabits the country beyond lake *Baikal*, and extends through the great *Gobèe*, even to *Thibet*. The *Tanguts* call it *Rangwo*, and confecrate it among the fpots of the moon* : agrees with the common rabbet in color of the flefh ; but does not burrow, running inftantly (without taking a ring as the common hare does) for fhelter, when purfued, into holes of rocks ; fo agrees in nature with neither that nor the rabbet. Called by the *Mongols*, *Tolai*. The fur is bad, and of no ufe in commerce.

Lepus Capenfis. L. cauda longitudine capitis, pedibus rubris. *Lin. fyft.* 78.

304. CAPE.

H with long ears dilated in the middle : the outfides naked, and of a rofe-color : infide and edges covered with fhort grey hairs : crown and back dufky, mixed with tawny : cheeks and fides cinereous : breaft, belly, and legs, ruft-colored : tail bufhy, carried upwards ; of a pale ferruginous color.

Size of a rabbet.

Inhabits the country three days north of the *Cape of Good Hope.* Is called there the *Mountain Hare*, for it lives only in the rocky mountains ; does not burrow. It is difficult to fhoot it, as it inftantly, on the fight of any one, runs into the fiffures of the rocks.

The fame fpecies probably extends as high as *Senegal*. M. *Adanfon* (44) obferves, that the hares of that country are not fo large

* *Pallas nov. fp.* i. 20.

as thofe of *France*; their color between that of the *European* kind
and a rabbet; and their flesh white.

---

305. Viscaccia.   Lepus vifcaccia. *Molina Chili,* 289. *Acofta*   *Peru.* 1725. p. 32. *Garcilaffo de la Ve-*
in *Purchas's* Pilgrims, iii. 966. *Feuillée*   *ga* 331.

H. with the appearance of a rabbet, excepting the tail; in that
part and color like a fox: the tail is long, and turned up,
and covered with coarfe hair: the reft of the hair foft: fize
fuperior to that of a rabbet.

Manners.   Inhabits *Peru* and *Chili:* lives under ground, and forms two
boroughs one above the other; in the one it keeps its provifions,
in the other fleeps: goes out only in the night: its flefh is white
and tender.   The antient *Peruvians* make ftuffs of the hair, which
were fo fine as to be worn only by the nobility.   In *Chili* it goes
into the hat-manufactory: its tail is its weapon of defence.

---

306. Cuy.   Lepus pufillus. *Molina Chili.* 288.

H. with a conoid body: ears fmall, pointed, and covered with
hair: nofe long: tail fo fhort as fcarcely to be feen: is
domefticated and varies in color to white, brown, and fpotted with
divers colors: fur very fine: fize of a field moufe.

Inhabits *Chili:* breeds every month, and brings from fix to
eight young: is delicate eating.

                                                        ** Without

** Without a tail.

Tapeti. *Marcgrave Brafil.* 223. *Pifo Brafil.* 102.
Cuniculus *Brafilienfis* Tapeti dictus. *Raii fyn. quad.* 205.
Lepus *Brafilienfis.* L. cauda nulla. *Lin.*

fyft. 78.
Lepus ecaudatus. *Briffon quad.* 97.
Le Tapeti. *De Buffon,* xv. 162.
Collar'd Rabbet. *Wafer's voy. in Dampier,* iii. 401.

307. BRASILIAN.

H. with very large ears, like the common kind: a white ring round the neck: face of a reddifh color: chin white: black eyes: color of the body like the common hare, only darker: belly whitifh: no tail: fome want the white ring round the neck.

Inhabit *Brafil:* live in woods: do not burrow: are very prolific: very good meat: found alfo in *Mexico**, where they are called *Citli.*

Lepus Alpinus. *Pallas,* nov. fp. fafc. i. 52. *tab.* ii. *Itin.* ii. 701. *tab.* A. *Zimmerman.*

308. ALPINE.

H. with fhort, broad, rounded ears: head long: very long whifkers: two very long hairs above each eye: color of the fur at the bottom dufky, towards the ends of a bright ferruginous; the tips white; intermixed are feveral long dufky hairs; but on firft infpection the whole feems of a bright bay.

Length of that I faw was nine inches.

* *Hernandez An. Nov. Hifp.* 2.

P 2

Thefe

PLACE.

Thefe animals are firft feen on the *Altaic* chain, and extend to lake *Baikal*; and from thence to *Kamtfchatka*; and, as it is faid, in the new-difcovered *Fox* or *Aleutian* iflands. They inhabit always the middle region of the fnowy mountains, in the rudeft places, wooded and abounding with herbs and moifture.

They fometimes form burrows between the rocks, and oftener lodge in the crevices; and are found in pairs, or more, according to conveniency: in cloudy weather they collect together, and

VOICE.

lie on the rocks, and give a keen whiftle, fo like that of a fparrow, as to deceive the hearer. On the report of a gun, they run into their holes; but foon come out again, fuppofing it to be a clap of thunder, to which they are fo much ufed in their lofty habitations.

By wonderful inftinct they make a provifion againft the rigorous feafon in their inclement feats. A company of them, towards autumn, collect together vaft heaps of choice herbs and graffes, nicely dried, which they place either beneath the overhanging rocks, or between the chafms, or round the trunk of fome tree. The way to thefe heaps is marked by a worn path. In many places the herbs appeared fcattered, as if to be dried in the fun and harvefted properly. The heaps are formed like round or conoid ricks; and are of various fizes, according to the number of the fociety employed in forming them. They are fometimes of a man's height, and many feet in diameter, but ufually about three feet.

Thus they wifely provide their winter's ftock, otherwife they muft perifh, being prevented by the depth of fnow to quit their retreats in queft of food.

8

They

They felect the beft of vegetables, and crop them when in the fulleft vigor, which they make into the beft and greeneft hay by the judicious manner in which they dry it. Thefe ricks are the origin of fertility amidft the rocks; for the reliques, mixed with the dung of the animals, rot in the barren chafms, and create a foil-productive of vegetables.

Thefe ricks are alfo of great fervice to that branch of mankind who devote themfelves to the laborious employ of fable-hunting: for being obliged to go far from home, their horfes would often perifh for want, if they had not the provifion of thefe induftrious little animals to fupport them; which is eafily to be difcovered by their height and form, even when covered with fnow. It is for this reafon that this little beaft has a name among every *Siberian* and *Tartarian* nation, which otherwife would have been overlooked and defpifed. The people of *Jakutz* are faid to feed both their horfes and cattle with the reliques of the winter ftock of thefe hares.

Thefe animals are neglected as a food by mankind, but are the prey of *fables* and the *Siberian* weefel, which are joint inhabitants of the mountains. They are likewife greatly infefted by a fort of *gadfly*, which lodges its egg in their fkin in *Auguft* and *September*, which often proves deftructive to them.

Lepus Ogotona. *Pallas* Nov. fp. fafc. i. 59. *tab.* iii.

309. OGOTONA.

H with oblong oval ears, a little pointed: fhorter whifkers than the former: hairs long and fmooth: color of thofe

on

on the body, brown at the roots, light grey in the middle, white at the ends, intermixed with a very few dusky hairs: a yellowish spot on the nose: space about the rump of the same color: outside of the limbs yellowish: belly white.

Length about six inches: weight of a male, from six ounces and a half to seven and a quarter; of the female, from four to four and three quarters.

Inhabits only the country beyond lake *Baikal,* and from thence common in all parts of the *Mongolian* desert, and the vast desert of *Gobee,* which extends on the back of *China* and *Thibet,* even to *India.* It lives in the open vallies, and on gravelly or rocky naked mountains. These little creatures are called by the *Mongols, Ogotona:* are found in vast abundance: live under heaps of stones, or burrow in the sandy soil, leaving two or three entrances. Their holes run obliquely: in these they make a nest of soft grass. The old females make for security many of these burrows near each other, that they may, if disturbed, retreat from one to the other.

They wander out chiefly in the night. Their voice is excessively shrill, a note like that of a sparrow, twice or thrice repeated; but very easily to be distinguished from that of the *Alpine* hare.

They live in the vallies, principally on the tender bark of a sort of *Service* and the dwarf elm; in the spring on different herbs. Before the approach of severe cold, in the early spring, they collect great quantities of herbs, and fill their holes with them, which the inhabitants of the country consider as a sure sign of change of weather.

Directed by the same instinct with the former, they form in

autumn

autumn their ricks of hay of a hemifpherical fhape; about a foot high and wide: in the fpring thefe elegant heaps difappear, and nothing but the reliques are feen.

They copulate in the fpring, and about the latter end of *June* their young are obferved to be full grown.

They are the prey of hawks, magpies, and owls: but the Cat *Manul* makes the greateft havock among them: and the ermine and fitchet is equally their enemy.

Lepus pufillus. *Pallas* Nov. fp. i. 31. *tab.* i. Nov. Com. *Petrop.* xiii. 531. *tab.* xiv.   **310. CALLING.**
Zimmerman.

H with a head longer than ufual with hares, and thickly covered with fur, even to the tip of the nofe: numerous hairs in the whifkers: ears large and rounded: legs very fhort: foles furred beneath: its whole coat very foft, long, and fmooth, with a thick long fine down beneath, of a brownifh lead-color: the hairs of the fame color; towards the ends of a light grey, and tipt with black: the lower part of the body hoary: the fides and ends of the fur yellowifh.

Length about fix inches: weight from three ounces and a   **SIZE.**
quarter to four and a half; in winter fcarcely two and a half.

Inhabits the fouth-eaft parts of *Ruffia*, and about all the ridge   **PLACE.**
of hills fpreading fouthward from the *Urallian* chain; alfo about the *Irtifh*, and in the weft part of the *Altaic* chain; but no
wh ere

H A R E.

where in the eaſt beyond the *Oby*. They delight in the moſt ſunny vallies, and herby hills, eſpecially near the edges of woods, to which they run on any alarm.

They live ſo concealed a life as very rarely to be ſeen: but are often taken in winter, in the ſnares laid for the ermines; ſo are well known to the hunters. About the *Volga* they are called *Semlanoi Saetſhik*, or *Ground Hare:* the *Tartars*, from their voice, ſtyle them *Tſchotſchot* or *Ittſitſkan*, or the *Barking Mouſe*; the *Kalmucs* call them *Ruſla*.

They chuſe for their burrows a dry ſpot, amidſt buſhes covered with a firm ſod, preferring the weſtern ſides of the hills; in theſe they burrow, leaving a very ſmall hole for the entrance; and forming long galleries, in which they make their neſts: but thoſe of the old ones, and females, are numerous and intricate: their place would be ſcarcely known but for their excrements, and even thoſe they drop, by a wiſe inſtinct, under ſome buſh, leaſt their dwelling ſhould be diſcovered by their enemies among the animal creation.

It is their voice alone that betrays their abode: it is like the piping of a quail, but deeper, and ſo loud as to be heard at the diſtance of half a *German* mile. It is repeated by juſt intervals thrice, four times, and often ſix. This is wonderful, as this little animal does not appear to be particularly organized for the purpoſe. The voice is emitted at night and morning: ſeldom in the day, except in rainy and cloudy weather. It is common to both ſexes; but the female is ſilent for ſome time after parturition, which is about the beginning of *May*, N. S. They bring forth ſix at a time, blind, and naked; which ſhe ſuckles often, and covers carefully with the materials of her neſt.

Theſe

1. *Calling Hare___N.310.*
2. *Alpine Hare___N.308.*
3. *Ogotona Hare___N.309.*

These most harmless and inoffensive animals never go from their holes: feed and make their little excursions by night: drink often: sleep little: are easily made tame: will scarcely bite when handled. The males in confinement are observed to attack one another, and express their anger by a grunting noise.

XXVII. BEAVER.

Two cutting teeth in each jaw.

Five toes on each foot.

Tail compreſſed, and covered with ſcales.

311. CASTOR.

Καστωρ. *Ariſt. hiſt. An. lib.* viii. *c.* 5. *Oppian. Halieut.* i. 398.
Fiber. *Plinii lib.* viii *c.* 30. *Agricola An. Subt.* 482. *Belon Aquat.* 25.
Caſtor. *Geſner quad.* 309. *Rondel.* 236. *Schoneveld Icth.* 34.
Beaver. *Raii ſyn. quad.* 209.
Bobr. *Rzaczinſki Polon.* 215.
Biber. *Klein quad.* 91. *Kramer Auſtr.* 315.

Caſtor caſtanei coloris, cauda horizonta-liter plana. *Briſſon quad.* 90.
Caſtor Fiber. C. cauda ovata plana. *Lin. ſyſt.* 78.
Baſwer, Biur. *Faun. ſuec.* N° 27.
Le Caſtor, ou Le Bievre. *De Buffon,* viii. 282. *tab.* xxxvi.
Beaver. *Br. Zool.* i. *Pl.* 9. LEV. MUS.

B. with ſtrong cutting teeth: ſhort ears, hid in the fur: blunt noſe: hair of a deep cheſnut brown: tail broad, almoſt oval, compreſſed horizontally, covered with ſcales: the fore feet ſmall; the hind large: length from noſe to tail, about three feet: tail eleven inches long, three broad.

PLACE.

Inhabits *Europe*, from *Lapland* to *Languedoc* * : in great plenty in the *North*: a few are yet found in the *Rhone* †, the *Gardon*, the *Danube*, the *Rhine*, and the *Viſtula*. I have an inſtance of two old and ſix young being taken in 1742, at *Gornichem*, in *Holland*; another in 1757 in the *Yſſel*, in *Guelderland*; and another in 1770 in the *Maas*, near the village *Hedel*, not far from *Bois le duc*: this laſt weighed forty pounds, and had two bags of *caſtoreum*, weigh-

* *De Buffon*, viii. 286.　　　† *Ibid.*

ing

Castor Beaver___N.311.

ing four ounces, and of excellent quality. It had inhabited the river for fome years, and done much damage to the willow-trees, with whofe bark its ftomach was found full. They are much more frequent in the *Lippe,* above *Wefel,* from which river they might defcend into thofe of *Holland* *.

Abound in the *Afiatic* part of the *Ruffian* empire; are found in companies, or affociated, about the *Konda,* and other rivers which flow into the *Oby.* They are met with difperfed, or in the ftate of *Terriers,* in the wooded parts of independent *Tartary,* and in the chains of mountains which border upon *Siberia.* None are to be feen in *Kamtfchatka,* by reafon of the interruption of the woods beyond the river *Kowyma;* nor yet in the new-difcovered iflands weft of that country: only in the ifle of *Kadjak,* the neareft to *America,* fome fkins have been procured by the *Ruffians,* which probably were got by the natives from *America,* in whofe northern parts they are found in prodigious abundance.

The moft induftrious of animals: nothing equals the art with which they conftruct their dwellings. They chufe a level piece of ground, with a fmall rivulet running through it. This they form into a pond, by making a dam acrofs; firft by driving into the ground ftakes five or fix feet long, placed in rows, wattling each row with pliant twigs, and filling the interftices with clay, ramming it down clofe. The fide neareft to the water is floped; the other perpendicular. The bottom is from ten to twelve feet thick; but the thicknefs gradually diminifhes to the top, which is about two or three. The length of thefe dams is fometimes not lefs than a hundred feet.

MANNERS.

* *Martine's Katechifm. Natur.* ii. 143.

Q 2

Their

Their houfes are made in the water collected by means of the dam, and are placed near the edge of the fhore. They are built on piles; are either round or oval; but the tops are vaulted; fo that their infide refembles an oven, the top a dome. The walls are two feet thick; made of earth, ftones, and fticks, moft artificially laid together; and the walls within as neatly plaiftered as if with a trowel. In each houfe are two openings; one into the water, the other towards the land. The height of thefe houfes above the water is eight feet. They often make two or three ftories in each dwelling, for the convenience of change, in cafe of floods. Each houfe contains from two to thirty beavers; and the number of houfes in each pond is from ten to twenty-five. Each beaver forms its bed of mofs; and each family forms its magazine of winter provifion, which confifts of bark and boughs of trees. This they lodge under water, and fetch it into their apartments as their wants require. *Lawfon* fays they are fondeft of the *faffafras*, afh, and fweet-gum. Their fummer food is leaves, fruits, and fometimes crabs and craw-fifh; but they are not fond of fifh.

To effect thefe works, a community of two or three hundred affembles; each bears his fhare in the labor: fome fall, by gnawing with their teeth, trees of great fize, to form beams or piles; thefe are gnawed all round in as regular a manner as a cutter cuts in falling a tree, bringing the bottom of the wood to a point *: others roll the pieces along to the water; others dive, and with their feet fcrape holes, in order to place them in; while others

* This will be beft underftood by infpecting the fpecimens in the LEVERIAN MUSEUM.

exert

exert their efforts to rear them in their proper places: another party is employed in collecting twigs, to wattle the piles with; a third, in collecting earth, ſtones, and clay; a fourth is buſied in beating and tempering the mortar; others, in carrying it on their broad tails to proper places, and with the ſame inſtrument ram it between the piles, or plaiſter the inſide of their houſes. A certain number of ſmart ſtrokes with their tail, is a ſignal given by the overſeer, for repairing to ſuch or ſuch places, either for mending any defects, or at the approach of an enemy; and the whole ſociety attend to it with the utmoſt aſſiduity. Their time of building is early in the ſummer; for in winter they never ſtir but to their magazines of proviſions, and during that ſeaſon are very fat. They breed once a year, and bring forth, the latter end of the winter, two or three young at a birth.

Beſides theſe aſſociated beavers, is another ſort, which are called *Terriers*; which either want induſtry or ſagacity to form houſes like the others. They burrow in the banks of rivers, making their holes beneath the freezing depth of the water, and work up for a great number of feet. Theſe alſo form their winter ſtock of proviſion.

Beavers vary in their colors: the fineſt are black; but the general color is a cheſnut brown, more or leſs dark: ſome have been found, but very rarely, white; others ſpotted: both varieties are preſerved in the Leverian Museum. The ſkins are a prodigious article of trade; being the foundation of the hat-manufactory. In 1763 were ſold, in a ſingle ſale of the *Hudſon's Bay* Company, 54,670 ſkins. They are diſtinguiſhed by different names. *Coat Beaver* is what has been worn as coverlets by the *Indians*. *Parchment Beaver,* becauſe the lower ſide reſembles it.

Stage

Stage Beaver is the worſt, and is that which the Indians kill out
of feafon, on their ſtages or journies. The valuable drug *Caſto-*
reum * is taken from the inguinal glands of theſe animals. The
antients had a notion it was lodged in the teſticles, and that the
animal, when hard preſſed, would bite them off, and leave them
to its purſuers, as if conſcious of what they wanted to deſtroy
him for.

> *Imitatus* Caſtora, *qui ſe*
> *Eunuchum ipſe facit, cupiens evadere damno*
> *Teſticulorum.*   JUVENAL, xii. 34.

> Juſt as the BEAVER, that wiſe thinking brute,
> Who, when hard hunted on a cloſe purſuit,
> Bites off the parts, the cauſe of all the ſtrife,
> And leaves them as a ranſom for his life.   DRYDEN.

---

* The *Ruſſian Caſtoreum* is ſo much better than the *American,* that we give two
guineas a pound for that, and only 8*s.* 6*d.* for the laſt; the firſt being leſs waxy,
and pulveriſes readier. Notwithſtanding we take this drug from *Ruſſia,* we export
there vaſt numbers of Beaver ſkins. The fleſh is reckoned good eating, being
preſerved, after the bones are taken out, by drying it in the ſmoke.

                                        MS. hiſt. Hudſon's Bay.

Muſſaſcus.

Muffafcus. *Smith's Virginia*, 27.
Mufquafh. *Joffelyn's voy. New England*, 86.
Mufk Rat. *Lawfon Carolina*, 120.
Caftor Zibethicus. C. cauda longa com-preffo-lanceolata, pedibus fiffis. *Lin. fyft.*

Caftor cauda verticaliter plana, digitis omnibus a fe invicem feparatis. *Briffon quad.* 93.
L'Ondatra. *De Buffon*, x. i. *tab.* i.
Rat Mufquè. *Charlevoix Nouv. France*, v. 157. *Lefcarbot N. Fr.* 350. LEV. MUS.

312. MUSK.

B. with a thick blunt nofe: ears fhort, and almoft hid in the fur: eyes large: toes on each foot feparated; thofe behind fringed on each fide with ftrong hairs, clofely fet together: tail compreffed fideways, and very thin at the edges, covered with fmall fcales, intermixed with a few hairs: color of the head and body a reddifh brown: breaft and belly afh-color, tinged with red: the fur very fine: length, from nofe to tail, one foot; of the tail, nine inches: in the form of its body, exactly refembles a beaver.

Inhabits *North America*: breeds three or four times in a year *, and brings from three to fix young at a time: during fummer, the male and female confort together: at approach of winter, unite in families, and retire into fmall round edifices, covered with a dome, formed of herbs and reeds cemented with clay: at the bottom are feveral pipes, through which they pafs in fearch of food; for they do not form magazines like the beavers: dur-ing winter, their habitations are covered many feet deep with fnow and ice; but they creep out and feed on the roots that lie

* MS. *hift. Hudfon's Bay.*

5

beneath:

beneath : they quit their old habitations annually, and build new ones. The fur is foft, and much efteemed : the whole animal, during fummer, has a moft exquifite mufky fmell : which it lofes in winter : perhaps the fcent is derived from the *Calamus Aromaticus*, a favorite food of this animal. *Lefcarbot* fays they are very good to eat.

313. GUILLINO.　　　　　　Caftor Huidobrius. *Molina Chili*, 266.

B. with a fquare head : fhort and round ears : fmall eyes : color grey ; dark on the back, whitifh on the belly. It has two forts of hair, like the common beaver : one fhort and fine, and fufceptible of any dye ; the other fpecies of hair long and hard : the toes of the fore feet bordered with a membrane ; the hind feet webbed : the back very broad : the tail long and hairy, and length from the nofe to the tail three feet ; height two feet.

MANNERS.　Inhabits the deepeft rivers and lakes of *Chili :* has the *foramen ovale* half clofed : can live long under water : feeds on fifhes and crabs : is fierce and bold, and will feize its prey in fight of mankind : is killed by the hunters when it comes to difcharge its excrements, which it does always in the fame place : moft beautiful ftuffs are made of the fur, refembling velvet ; it is alfo of great ufe in the manufacture of hats.

M. *Molina* calls it *Huidobrius*, from the family name of his patron, the marquifs of *Cafa Reale*.

M. *Molina*

M. *Molina* was one of the *Jefuits* whom the *Spaniards* expelled out of *South America*. They robbed him of all his effects and manufcripts: by a fingular fortune he found in *Italy* the manufcript which furnifhes us with the valuable natural hiftory of *Chili*.

XXVIII.
PORCUPINE.

Two cutting teeth in each jaw.
Body covered with long, hard, and sharp quills.
Upper lip divided.

314. CRESTED.

Υϛδιξ. *Ariſtot. hiſt. An. lib.* i. *c.* 6. *Oppian Cyneg.* iii. 391.
Hyſtrix. *Plinii lib.* viii. *c.* 35. *Geſner quad.* 563. *Raii ſyn. quad.* 206.
Acanthion criſtatus. *Klein quad.* 66.
Hyſtrix orientalis criſtata. *Seb. Muſ.* i. 79. *tab.* l.

Hyſtrix criſtata. H. palmis tetradactylis, plantis pentadactylis, capite criſtato, cauda abbreviata. *Lin. ſyſt.* 76. *Haſſelquiſt. itin.* 290.
Hyſtrix capite criſtato. *Briſſon quad.* 85.
Le Porc-epic. *De Buffon,* xii. 402. *tab.* li. lii. *Faunul. Sinens.*

P. with a long creſt on the top of the head, reclining backwards, formed of ſtiff briſtles : the body covered with long quills; thoſe on the hind part of the body nine inches in length, very ſharp at the ends, varied with black and white; between the quills a few hairs : the head, belly, and legs, are covered with ſtrong briſtles, terminated with ſoft hair, of a duſky color : the whiſkers long : ears like the human : four toes before, five behind : tail ſhort, and covered with quills : length, from noſe to tail, two feet; tail, four inches.

Inhabits *India,* the ſand-hills on the S. W. of the *Caſpian* ſea, ſouthern *Tartary, Perſia,* and *Paleſtine,* and all parts of *Africa* : is found wild in *Italy*; but is not originally a native of * *Europe* : is brought into the markets of *Rome,* where it is eat †. The *Italian* porcupines have ſhorter quills, and a leſſer creſt, than thoſe of *Aſia* and *Africa* : is an harmleſs animal : lives on fruits, roots, and vegetables : ſleeps by day, feeds by night : the report of its

* *Agricola An. Subt.* 486.
† *Ray's Travels,* i. 311. *Ph. Tr. abridg.* v. 147.

darting

Long-tail'd Porcupine _ N.º 316.

darting its quills fabulous: when angry, retires and runs its nose into a corner, erects its spines, and opposes them to its assailant: makes a snorting noise.

These animals produce a *Bezoar*; but, according to *Seba*, only those which inhabit *Java*, *Sumatra*, and *Malacca*. These *Bezoars* were very highly valued, and have been sold for five hundred crowns apiece. It has also been pretended that a stone was procured from the head of this animal, infinitely more efficacious than other *Bezoars** ; but this may be placed among the many impositions of oriental empirics.

Erinaceus Malacensis. *Gm. Lin.* 116. *Seb. Mus.* i. p. 81. tab. 41. fig. i.   315. MALACCA.

P. with large pendulous ears: no crest: quills like the preceding, with the interstices filled with long hairs, resembling bristles: eyes large and bright: hair on the legs, and belly covered with short reddish prickly hairs: toes five in number, which might determine *Linnæus* to place this animal among the hedge-hogs.

Inhabits the peninsula of *Malacca*.

Porcus aculeatus sylvestris, seu Hystrix orientalis singularis. *Seb. Mus.* i. 84. tab. lii.
Acanthion cauda prælonga, acutis pilis horrida, in exitu quasi panniculata. *Klein quad.* 67.

Hystrix cauda longissima, aculeis undique obsita, in extremo panniculata. *Brisson quad.* 89.
Hystrix macroura. H. pedibus pentadactylis, cauda longissima: aculeis clavatis. *Lin. syst.* 77.

316. LONG-TAILED.

P. with long whiskers: short naked ears: large bright eyes: body short and thick, covered with long stiff hairs as sharp

* *Tavernier*, ii. 154.

R 2   as

as needles, of different colors, according as the rays of light fall on them: feet divided into five toes; that which ferves as a thumb turns backwards: the tail is as long as the body, very flender to the end, which confifts of a thick tuft: the briftles appearing as if jointed; are thick in the middle, and rife one out of the other like grains of rice; are tranfparent, and of a filvery appearance.

Inhabits the ifles of the *Indian Archipelago*, and lives in the forefts.

317. BRASILIAN.

Tlaquatzin. *Hernandez, Mex.* 330.
Cuandu. *Brafilienfibus, Lufitanis.*
Ourico cachiero. *Marcgrave Brafil.* 233.
   Pifo Brafil. 99. 325.
Iron Pig. *Nieuhoff,* 17.
Hyftrix Americanus. *Raii fyn. quad.* 208.
Hyftrix prehenfilis, H. pedibus tetradac-tylis, cauda elongata prehenfili femi-nuda. *Lin. fyft.* 76.

H. cauda longiffima, tenui, medietate ex-trema aculeorum experte, 87.
H. Americanus major, 88.
Hyftrix longius caudatus, brevioribus aculeis. *Barrere France Æquin.* 153.
Hyftrix minor leucophæus, Gouandou. *ibid.*
Chat epineux. *Des Marchais,* iii. 303.

P. with a fhort blunt nofe: long white whifkers: beneath the nofe a bed of fmall fpines: top of the head, back, fides, and bafe of the tail, covered with fpines; the longeft, on the lower part of the back and tail, are three inches in length, very fharp, white, barred near their points with black; adhere clofely to the fkin, which is quite naked between them; are fhorter and weaker as they approach the belly: on the breaft, belly, and lower part of the legs, are converted into dark-brown briftles: feet divided into four toes: claws very long; on the place of the thumb a great protuberance: tail eighteen inches long, flender, and taper towards the end; the laft ten inches is almoft naked, having

only

Brasilian Porcupine ___ N.°317.

only a few hairs on it; has, for that length, a ftrong prehenfile quality.

Inhabits *Mexico* and *Brafil*: and extends to *Chili*: lives in the woods: preys not only on fruits, but poultry: fleeps in the day, preys by night: makes a noife with its noftrils as if out of breath: grunts like a fow*: grows very fat: its flefh white, and very good: climbs trees, but very flowly; in defcending, twifts its tail round the branches, for fear of falling: is no more capable of fhooting its quills than the firft: may be tamed. *Pifo* fays there is a greater and leffer kind.

This fpecies is very rarely brought into *Europe*. I had opportunity of defcribing it from a fpecimen fome time in poffeffion of Mr. *Greenwood*; who was fo obliging as to permit me to have a drawing made of it, from which a very faithful figure is here given. M. *de Buffon* † has made mention of this animal in his work; but unjuftly reproaches *Marcgrave* with confounding it with the *Mexican* fpecies.

Hoitzlacuatzin, feu Tlacuatzin fpinofus, Hyftrix novæ Hifpaniæ. *Hernandez Mex.* 322.
Hyftrix novæ Hifpaniæ. H. aculeis apparentibus, cauda brevi et craffo. *Briffon quad.* 86.
Le Coendu. *De Buffon*, xii. 421. *tab.* liv.

318. MEXICAN.

P. of a dufky color, with very long briftles intermixed with the down: the fpines three inches long, flender, and varied with white and yellow; fcarcely apparent, except on the tail,

* *Vocem edit ut Sus*, iii. Marcgrave, 233.
† Under the name of *Le Coendou*, xii. 421. *tab.* liv.

which

which is, according to *Hernandez*, thicker and shorter than that of the preceding species. He adds, that the tail, from the middle to the end, is free from spines.

SIZE.

According to *Hernandez*, it grows to the bulk of a middle-sized dog. M. *de Buffon* says, its length is sixteen or seventeen inches from the nose to the tail; the tail nine, *French* measure, but taken from a mutilated skin.

PLACE.

Inhabits the mountains of *Mexico*: lives on the summer fruits, and may be easily made tame. The *Indians* pulverise the quills, and say they are very efficacious in gravelly cases; and, applied whole to the forehead, will relieve the most violent head-ach. They adhere till filled with blood, and then drop off.

319. CANADA.

Porcupine from *Hudson's Bay*. *Edw*. 52. *Ellis's voy*. 42. *Clerk's voy*. i. 177. 191. Cavia Hudsonis. *Klein quad*. 51. Hystrix dorsata. H. palmis tetradactylis, plantis pentadactylis, cauda mediocri, dorso solo spinoso. *Lin. syst*. 76. Hystrix aculeis sub pilis occultis, cauda brevi et crassa. *Brisson quad*. 87. L'Urson. *De Buffon*, xii. 426. *tab*. lv. LEV. Mus.

P. with short ears, hid in the fur: head, body, legs, and upper part of the tail, covered with soft, long, dark brown hair: on the upper part of the head, back, body, and tail, numbers of sharp strong quills; the longest on the back, the left towards the head and sides; the longest three inches; but all are hid in the hair: intermixed, are some stiff straggling hairs, three inches longer than the rest, tipt with dirty white: under side of the tail white: four toes on the fore feet, five behind, each armed with long claws, hollowed on their under side: the form of the body is exactly that of a beaver; but is not half the size:

3

one,

one, which Mr. *Banks* brought from *Newfoundland*, was about the fize of a hare, but more compactly made: the tail about fix' inches long.

Thefe animals vary in color. Sir *Afhton Lever* had one, which is entirely white.

Inhabits *N. America*, as high as *Hudfon's Bay*: makes its neft under the roots of great trees, and will alfo climb among the boughs, which the *Indians* cut down when one is in them, and kill the animal by ftriking it over the nofe: are very plentiful near *Hudfon's Bay*, and many of the trading *Indians* depend on them for food, efteeming them both wholefome and pleafant: feed on wild fruits and bark of trees, efpecially juniper: eat fnow in winter, drink water in fummer; but avoid going into it: when they cannot avoid their purfuer, will fidle towards him, in order to touch him with the quills, which feem but weak weapons of offence; for, on ftroking the hair, they will come out of the fkin, fticking to the hand. The *Indians* ftick them in their nofes and ears, to make holes for the placing their ear-rings and other finery: they alfo trim the edges of their deer-fkin habits with fringes made of the quills, or cover with them their bark-boxes.

PLACE.

Two

Two cutting teeth in each jaw.

Four toes before, five behind.

Short ears, or none.

Tail covered with hair, and of a middling length; in some very short.

320. ALPINE.

Mus Alpinus. *Plinii lib.* viii. *c.* 37. *Agricola An. Subter.* 484. *Gefner quad.* 743. *Raii fyn. quad.* 221.
Glis marmota. *Klein quad.* 56. *Hift. Mur. Alp.* 230.
Murmelthier. *Kramer Auftr.* 317.
Mus marmota. M. cauda abbreviata fub-pilofa, auriculis-rotundatis, buccis gibbis. *Lin. fyft.* 81.
Glis pilis e fufco et flavicante mixtis veftitus. Glis flavicans, capite rufefcente. *Briffon quad.* 116, 117.
La Marmotte. *De Buffon*, viii. 219. *tab.* xxviii.

M. with short round ears, hid in the fur: cheeks large: color of the head and upper part of the body brownish ash, mixt with tawny: legs and lower part of the body reddish: tail pretty full of hair: length, from nofe to tail, about sixteen inches; tail six: body thick.

PLACE.

Inhabits the loftiest summits of the *Alps* and *Pyrenæan* mountains: feeds on infects, roots, and vegetables: while they are at food, place a centinel, who gives a whistle on feeing any sign of danger, on which they instantly retire into their holes: form holes under ground, with three chambers of the shape of a Y, with two entrances; line them well with mofs and hay; retire into them about *Michaelmas*, and, stopping up the entrances with earth, continue in a torpid state till *April*: when taken out remain infenfible, except brought before a fire, which revives

5   them:

them: they lodge in fociety, from five to a dozen in a chamber: will walk on their hind feet: lift up their meat to their mouth with their fore feet, and eat it fitting up: bring three or four young at a time: are very playful: when angry, or before a ftorm, make a moft ftrange noife; a whiftle fo loud and fo acute, as quite to pierce the ear: grow very fat about the backs: are fometimes eaten; but generally taken in order to be fhewn, efpecially by the *Savoyards*: grow very foon tame, and will then eat any thing: are very fond of milk, which they lap, making at the fame time a murmuring noife, expreffive of their fatisfaction: very apt to gnaw any cloaths or linen they find: will bite very hard.

M. with a blunt nofe: fhort rounded ears: cheeks puffed, and of a grey color: face dufky: nofe black: hair on the back grey at bottom, black in the middle, and the tips whitifh: belly and legs of an orange-color: toes black, naked, and quite divided; four, and the rudiments of another, on the fore feet; five behind: tail fhort, and of a dufky color: was rather larger than a rabbet.

321. QUEBEC.

Inhabits *Hudfon's Bay* and *Canada*. Mr. *Brooks* had one alive a few years ago; it was very tame, and made a hiffing noife: perhaps is the fpecies which the *French* of *Canada* call *Siffleur*.

PLACE.

It has lately been defcribed by Dr. *Pallas*, under the name of *Mus empetra* *.

* *Nov. fp. quadr. fafc.* i. 75.

322. MARYLAND.

Bahama Cony. *Catesby Carolina*, ii. 79.
Monax, *Catesby Carolina App.* xxviii.
Monax, or Marmotte of *America*, *Edw.*
    104.
Glis Marmota, Americanus. *Klein quad.*
    56. *De Buffon*, Suppl. iii. 175.

Glis fufcus. Glis fufcus, roftro e cinereo
    cærulefcente. *Briffon quad.* 115.
Mus Monax. M. cauda mediocri pilofa,
    corpore cinereo, auriculis fubrotundis,
    palmis tetradactylis, plantis pentadacty-
    lis. *Lin. fyft.* 81.

M with fhort rounded ears: black prominent eyes: nofe fharper than that of the laft: nofe and cheeks of a blueifh afh-color: back of a deep brown color: fides and belly paler: tail half the length of the body, covered with pretty long dufky hair: toes divided, and armed with fharp claws: four toes before, five behind: feet and legs black: is about the fize of a rabbet.

PLACE.   Inhabits *Virginia* and *Penfylvania*: during winter fleeps under the hollow roots of trees: is found alfo in the *Bahama* ifles: lives on wild fruits and other vegetables: its flefh is very good, tafting like that of a pig: when furprized, retreats to holes in the rocks: whether it fleeps, during winter, in the climate of thofe ifles, is not mentioned.

323. HOARY.

M with the tip of the nofe black: ears fhort and oval: cheeks whitifh: crown dufky and tawny: hair univerfally rude and long; that on the back, fides, and belly, cinereous at the root, black in the middle, whitifh at the tip, fo that the animal has a hoary appearance: legs black: claws dufky; four before, five behind: tail black, mixed with ruft-color.

About the fize of the former.

Inhabits

Inhabits the northern parts of *North America*. Defcribed from a fpecimen in the LEVERIAN MUSEUM.

Bobak Swiftch. *Rzaczinfki Polon.* 233.
Bobak. *Beauplan hift. Ukrain, Churchill's coll.* i. 6oo. *Forfter hift. Volgæ, Phil. Tranf.* lvii. 343. *De Buffon,* xiii. 136.

·tab. xviii.
Sogur. *Rubruquis's Travels in Purchas.* iii. 6.
Arctomys. *Pallas nov. fp. fafc.* i, 9. tab. v.

324. BOBAK.

M. with fmall oval thick ears, covered with greyifh white down; with longifh hairs on the edges: eyes fmall: whifkers fmall: color about the eyes and nofe dufky brown; among the whifkers ferruginous: upper part of the body greyifh, intermixed with long black or dufky hairs, tipt with grey: throat ruft-colored: reft of the body, and the infide of the limbs, of a yellowifh ruft-color: four toes on the fore feet, with a fhort thumb furnifhed with a ftrong claw: five toes behind: tail fhort, flender, full of hair.

Length from nofe to tail fixteen inches; of the trunk of the tail, about four: the hairs extend an inch beyond the end of the trunk.

SIZE.

Inhabits the high but milder and funny fides of mountanous countries, which abound with fiffile or free-ftone rocks: feek dry fituations, and fuch which are full of fprings, woods, or fand. They are found in *Poland*, and the fouth of *Ruffia*, among the *Carpathian* hills; they fwarm in the *Ukraine*, about the *Borifthenes*, efpecially between the *Sula* and *Supoy*; and again between the *Borifthenes* and the *Don*, and along the range of hills which extend to the *Volga*; they are found about the *Yaik* and other

PLACE.

neighboring

neighboring rivers. Inhabit the fouthern defert in *Great Tartary*, and the *Altaic* mountains eaft of the *Irtis*; ceafe to appear in *Siberia*, on account of its northern fituation; but are found again beyond lake *Baikal*, and about the river *Argun* and lake *Dalay*; in the funny mountains about the *Lena*; and very common in *Kamtfchatka*, but rarely reach as high as *lat.* 55.

MANNERS.

They burrow extremely deep, and obliquely, to the depth of two, three, or four yards: they form numbers of galleries with one common entrance from the furface; each gallery ends in the neft of the inhabitant. Sometimes the burrows confift of only one paffage. They are found in great abundance about the fepulchral tumuli, as they find they can penetrate with great facility in the foft dry earth; but they are very common in the rocky ftrata; and in the mineral part of the *Urallian* chain, often direct the miners to the veins of copper, by the fragments which appear at the mouth of their holes, flung out in the courfe of their labors. In very hard and rocky places, from twenty to forty of thefe animals join together to facilitate the work, and live in fociety, each with its neft at the end of its refpective gallery; but the feweft galleries are found in the fofteft ground, and very frequently only a fingle one. In each neft they collect, efpecially towards autumn, the fineft of hay, and in fuch plenty, that fufficient is found in one neft for a night's food for a horfe.

During the middle and funny part of the day they fport about the entrance of their holes, but feldom go far from them; on the fight of man they retire with a flow pace, and fit upright near the mouth, and give a frequent whiftle, liftening at the approach. In places where they live in large families, they always

ways

ways place a centinel to give notice of any danger, during the time the reft are feeding.

They are very fond of oleraceous plants: in a ftate of confinement eat cabbage and bread very greedily, and drink milk with great eagernefs; but refufe water, and feem never affected with thirft: they are mild and good-natured; never quarrel or fight about their food in a wild ftate, and when confined, and placed with others, caught in diftant parts, and ftrangers to them, grow inftantly familiar with them: then very foon become tame, even when taken in full age; but the young immediately become familiar.

The number produced at a birth is not certainly known, probably at times eight; the females being furnifhed with that number of teats: they breed early, for in *June* the young are obferved to be of half the fize of the old.

They lie torpid during winter, except thofe which are kept tame in the ftove-warmed rooms of the country; and even then, finding a defect of that warmth which the fnug neft of their fubterraneous retreat would afford, in cold nights creep for fhelter into the very beds of the inhabitants. In that ftate they will not abfolutely refufe food, but eat very little, and that with a feeming difguft; nature allotting for them, in the wild ftate, a long fleep and ceffation from food, the refult of plenitude previous to its commencement. They fometimes efcape from confinement, find a retreat, and get their winter's fleep, and return to their mafter in the fpring; but lofe much of their gentle manners.

They grow very fat: the fat is ufed for foftening of leather: the fkins are ufed by the *Koreki*, people of *Jakutks*, and the *Ruf-*
fians,

fians, for cloathing. The *Calmucs* take them in fmall nets with large mefhes, placed before their holes. The inhabitants of *Ukraine* catch them in *May* or *June,* by pouring water into the holes, which forces them into the nets. In *South Ruffia* they are deftroyed by means of a log of wood with a weight at top; the end directed into a wooden box placed at the mouth of the hole, which falls as foon as the animal comes out, and oppreffes it by the weight. Their flefh taftes like that of a hare, but is rank.

The *Calmucs* are very fond of the fat ones, and even efteem them medicinally: on the contrary, the *Mahometan Tartars* not only abftain from their flefh, but even give them protection; fo that near the hords they are extremely numerous: thefe *Tartars* efteem a warren of *Bobaks* near them to be very fortunate, and think it a fin to kill one of them, a fwallow, or a dove; but at the fame time abominate the following animal.

In *Chinefe Tartary* they are the propagators of *Rhubarb,* which grows among their burrows: the manure which they leave about the roots contributes to its increafe; and the loofe foil they fling up, proves a bed for the ripe feed; which, if fcattered among the long grafs, perifhes without ever being able to reach the ground.

Mus

1. Quebec Marmot___ N.321.

2. & 3. Earless Marmot___ N.326.

Mus Maulinus. *Molina Chili*, 284.

M. with pointed ears: elongated nose: whiskers disposed in four rows: the tail longer than that of the common kind: five toes on each foot; an anomalous distinction: hair like the common: in size twice as large.

Discovered in the province of *Maule* in *Chili*, in 1764, and inhabits the woods: makes a stout defence against the dogs, which conquer it not without difficulty.

MANNERS.

Mus Noricus aut Citellus. *Agricola Ar. Subter.* 485. *Gesner quad.* 737. *Raii syn. quad.* 220.

Ziesel. *Schwenkfelt. Theriotroph.* 86.

Mus citellus. M. cauda abbreviata, corpore cinereo, auriculis nullis. *Lin. syst.* 80.

Tsitsjan. *Le Bruyn voy. Musc.* ii. 402 *.

Cuniculus caudatus, auriculis nullis, cinereus. *Brisson quad.* 101.

Le Zisel. *De Buffon*, xv. 139.

Le Souslik —— 144. 195. *Supplem.* iii. 191. *tab.* xxxi.

Mus Marmotta. *Forster hist. nat. Volgæ.* Ph. Transf. lvii. 343.

Mus Citillus. *Pallas nov. sp. fasc.* i. 119. *tab.* vi. vii. B. *Nov. com. Petrop.* xiv. 549. *tab.* vii.

Earless Marmot. *Syn. quad.* 276. *Casan. M.* —— 272.

326. EARLESS.

M. with a cinereous face: over each eye a white line: teeth yellow: whiskers black and long: no ears: hind part of

* Un chien courant que j'avois, y prit dans la plaine un petit animal nommé *Zits-jan*, qu'il m'apporta en vie, et un autre peu après, lesquels je fis 'eventrer, pour les conserver. C'est un espece de rat de campagne, de la grosseur d'un écureuil, qui a la queuë courte, et le poil et la couleur d'un lapreau, aussi bien que la forme, hors qu'il a la tête plus grosse, et les deux dents de dessous la moitié plus longues que celles de dessus. Il a aussi les pattes de devant plus courtes que celles de derriere, avec quatre grifes, et une plus petite, et cinque à celles de derriere, ressemblant assez à celles d'un singe.

8

the

MARMOT.

the head, and whole back, of a pale yellowiſh brown; often diſtinctly ſpotted with white; ſometimes undulated with grey: under ſide of the body, and legs, of a yellowiſh white.

Tail covered with long hair; brown above, bordered with black on each ſide; each hair tipped with white: under part of a bright ruſt-color: three middle toes of the fore feet long: claws long and ſharp: exterior and interior toes ſhort; the laſt remote from the others: its claws ſhort and blunt.

ſIZE.

Length one foot; of the tail, to the end of the hairs, four inches and a half.

Inhabits *Bohemia, Auſtria, Hungary,* and from the banks of the *Volga* to *India* and *Perſia*; through *Siberia,* and *Great Tartary,* to *Kamtſchatka* *; ſome of the intervening iſles, ſuch as *Kadjak*; and even the continent of *America* itſelf.

Burrows, and forms its magazine of corn, nuts, &c. for its winter food † : ſits up like a ſquirrel while it eats: ſome inhabit the fields in *Siberia,* others penetrate into the granaries; the firſt form holes under ground, with a double entrance, where they ſleep during winter: thoſe which inhabit granaries, are in motion during the cold ſeaſon. They couple the beginning of *May,* about the *Lena,* but about *Aſtracan* earlier, and bring from five to eight young, which they bring up in their burrows, and cover with hay: only one animal inhabits each burrow: the females are always ſeparate from the males, except in the coupling ſeaſon: whiſtle like the marmot: are very iraſcible; quarrelſome among themſelves, and bite very hard: ſit in multitudes near their holes: are very fond of ſalt: taken in numbers on board the barges

---

* Yevraſhka, or Marmotte minor. *Gmelin voy. Siberia,* ii. 448.
† *Raii ſyn. quad.* 220.

which

which are loaden with that commodity at *Solikamſky*, and fall down into the *Volga* below *Caſan*.

Are both herbivorous and carnivorous; feed on plants, and deſtroy the young of ſmall birds, and the leſſer mice.

The *Bohemian* ladies were wont to make cloaks of the ſkins; we ſee them at this time made uſe of for linings, and appear very beautiful for that purpoſe, eſpecially the ſpotted kind.

M. with truncated ears, the apertures large: ſhort tail: upper fore teeth truncated; lower, ſlender and pointed: four toes on every foot, each furniſhed with claws: walks on the whole hind feet as far as the heel: color, teſtaceous red.

327. Gundi.

Size of a ſmall rabbet.

Inhabits *Barbary* towards *Mount Atlas*, near *Maſſuſin*. Deſcribed by the late Mr. *Rohtman*, a *Swede*. This account was communicated to me by Mr. *Zimmerman*. *Gundi* is its *Arabic* name, which I retain.

M. with ſhort ears: head and body of a cinereous brown; the ends of the hairs white: two cutting teeth above; four below: no tail.

328. Tailless.

I communicated a drawing of this ſpecies to Mr. *Bewick*, who has given an engraving of it at p. 374 of his ingenious performance. Inhabits *Hudſon's Bay*. In the Leverian Museum.

Vol. II.   T   With

XXX.
SQUIRREL.

With two cutting teeth in each jaw.
Four toes before, five behind.
Long tail, cloathed with long hair.

329. COMMON.

Sciurus. *Gefner quad.* 845. *Raii fyn. quad.* 214. — tadactylis. *Lin. fyft.* 86.
Wiewiorka. *Rzaczinfki Polon.* 225. — Ikorn, Grafkin *Faun. fucc.* N° 37.
Eichhorn. *Klein quad.* 53. — Sciurus rufus quandoque grifeo admixto. *Briffon quad.* 104.
Sciurus vulgaris. Sc. auiculis apice barbatis, pumis tetradactylis, plantis pen- — L'Écureuil. *De Buffon,* vii. 258. *tab.* xxxii. *Br. Zool.* i. 93. LEV. MUS.

S. with ears terminated with long tufts of hair: large lively black eyes: head, body, legs, and tail, of a bright reddifh brown: breaft and belly white: hair on each fide the tail lies flat. In *Sweden*, and *Lapland** changes in winter into grey. In many parts of *England* is a beautiful variety with milk-white tails.

Inhabits *Europe*; the northern and temperate parts of *Afia*; and a variety is even found as far *fouth* as the ifle of *Ceylon*: is a neat, lively, active animal: lives always in woods: in the fpring, the female is feen purfued from tree to tree by the males, feigning an efcape from their embraces. Makes its neft of mofs and dried leaves, between the fork of two branches: brings three or four young at a time: has two holes to its neft: ftops up that on the fide the wind blows, as *Pliny* + juftly remarks: lays in a hoard of

* *Faun. Suec.* and *Scheffer Lapl.* 135.   + *Lib.* viii. *c.* 38.

8   winter

winter provifion, fuch as nuts, acorns, &c.; in fummer, feeds on buds and young fhoots: is particularly fond of thofe of fir, and the young cones: fits up to eat, and ufes its fore-feet as hands: covers itfelf with its tail: leaps to a furprifing diftance: when difpofed to crofs a river, a piece of bark is its boat; its tail the fail *.

A large kind of grey fquirrel is found about the upper parts of the river *Obi*, in the diftrict of *Kuznetfk*, and is called *Teleutfkaya Belka*, or the fquirrel of the *Teleutian Tartars:* it is as large again as the common grey fquirrels of thofe parts, and is preferred to them, on account of the filvery glofs of the fkin. Few are fent into *Ruffia*, the greateft part being fent into *China*, and fell for 6 *l.* or 7 *l. fterling per* thoufand †.

A white variety is found common in *Siberia*.

A beautiful black variety about lake *Baikal*. In the LEVE-RIAN MUSEUM is a moft elegant fpecimen of this kind.

α. WHITE-LEGGED SQUIRREL. The head, whole upper part of the body, fides, and toes, of a reddifh brown: face, nofe, under fide of the neck, belly, fore legs, infide of the ears and thighs, white: ears flightly tufted with black: tail long, covered with dufky hairs, much fhorter than thofe in the *European* kind. *Br. Muf :* by the catalogue, faid to be brought from *Ceylon*.

---

* *Rzaczinfki, Klein, Scheffer, Linnæus.*
† *Memorabalia Ruff. Afiat. in Muller's Samlung. Ruff.* vii. 124.

330. CEYLON.   Sciurus *Zeylanicus*, pilis in dorfo nigrican-   Sciurus macrourus, long-tailed Squirrel.
                tibus, *Rukkaia* dictus a fono. *Raii fyn.*   *Ind. Zool. tab.* i.
                quad. 215.

S. with ears tufted with black: nofe flefh-colored: cheeks, legs, and belly, of a pale yellow: between the ears a yellow fpot: forehead, back, fides, haunches, black: cheeks marked with a bifurcated ftroke of black; under fide red: tail twice as long as the body, of a light grey, and very bufhy: the part next the body quite furrounded with hair: on the reft the hairs are feparated, and lie flat. Is thrice the fize of the *European* fquirrel.

Inhabits *Ceylon*: is called there *Dandoelana*: alfo *Roekea*, from the noife it makes.

331. ABYSSINIAN.   S. with a round flefh-colored nofe: hair on the upper part of the body of a rufty black: tail a foot and a half long: belly and fore feet grey: foles of the feet flefh-colored. Thrice the fize of an *European* fquirrel.

Defcribed from *Thevenot*[*], who fays it was bought at *Moco* from an *Abyffinian*; that it was very good-natured, and fportive like a fquirrel; would eat any thing except flefh, and would crack the hardeft almonds. A variety of the above?

[*] *Voyage des Indes Orientales*, v. 34.

Sciurus

Sciurus maximus. *Gmelin Lin.* i. 149. Grand Ecureuil. *Sonnerat, voy.* ii. 139.   332. MALABAR.

S. with fhort tufted ears: five toes to each foot: inftead of a thumb to the hind foot, is a fhort excrefcence, with a flat nail; all the other nails ftrong and crooked: tail very full of hair, and as long as the body: hair long, of a reddifh color, reflecting gold; a beard of the fame begins under each ear, and turns towards the body: all the hind part of the body and tail black: is of the fize of a cat.

Inhabits the mountains of *Cardomone* which form part of the   PLACE.
Gauts: is very fond of the milk of the coco nut, which it will pierce and fuck out on the tree. Its cry is fharp and piercing.

Sonnerat voy. ii. 140.   333. GINGI.

S. of a dirty grey color; brighteft on the belly: eyes encompaff- ed with a white circle: on each fide of the belly is a white line which extends along the fhoulders and thighs: tail black: rather larger than the *European* kind.

Inhabits *Gingi.*

Sonnerat,

334. AYE. AYE.   *Sonnerat*, ii. 142. tab. lxxxviii.

S. with large broad ears, fmooth, fhining, and with feveral long hairs fcattered over them: fur foft and fine; of a tawny white, intermixed with fome long black hairs: the tail is very buſhy, covered with long hairs, black at their ends, white at their bottoms: five toes to each foot: the two joints of the middle finger of the fore feet very flender.; the thumb of the hind foot furniſhed with a flat nail.

SIZE.

Length eighteen inches; tail of the fame length: burrows under ground: goes out only in the night: the eyes fixed: is very flothful, and of gentle manners: very fearful: much inclined to fleep; and refts with its head between its legs.

PLACE.

Inhabits *Madagafcar:* is a very rare animal: takes its name from its cry, the note of aftoniſhment of the natives of that iſland.

335. JAVAN.

S. black on the upper part of the body; of a light brown on the lower: end of the tail black: on the thumb a round nail.

This brief account leaves me uncertain whether this is not alfo a variety.

Inhabits *Java:* difcovered by Mr. *Sparman.* Memoirs fociety at *Gothenburgh.* Dr. PALLAS.

S. with

No. 334.

S with tufted ears: head, back, fides, upper part of the legs
• and thighs, and tail, of a dull purple: the lower part of
the legs, and thighs, and the belly, yellow: end of the tail orange:
length, from nofe to tail, near fixteen inches; tail feventeen.

336. BOMBAY.

Inhabits *Bombay*. Defcribed from a ftuffed fkin in Doctor *Hunter*'s cabinet.

This fpecies extends to *Balifere*, the oppofite part of the
peninfula of *Indoftan*.

M. *de la Cepede* * gives the defcription and figure of a large
fquirrel which agrees fo much with this, that I fufpect it to be
only a variety. He fays on one front of the face is a chefnut
fpot, furrounded with orange: in other refpects, there is much
agreement, only he makes no mention of the orange at the end of
the tail.

Sciurus Erythræus. *Pallas Nov. fp. fafc.* i. 377. *Miller's* plates. tab. xlvi.

337. RUDDY.

S with the ears flightly tufted: color above yellow, mixed
• with dufky: below of a blood red inclining to tawny: tail
flender; of the fame color, marked lengthways with a black
ftripe.

* See M. *de Buffon*, Suppl. vi. 254. tab. lxii.

Four

Four toes on the fore feet; with a remarkable protuberance inftead of a thumb: five toes on the hind.

Rather larger in fize than a common fquirrel.

Inhabits *India*.

338. GREY.

Grey Squirrel. *Joffelyn's voy. Catefby Carolina*, ii. 74. *Smith's voy.* 27. *Kalm's voy.* 95, 310.

Fox Squirrel. *Lawfon's Carolina*, 124.

Sciurus cinereus Virginianus major. *Raii fyn. quad.* 215.

Sciurus cinereus. *Lin. fyft.* 86.

Sciurus cinereus. Auriculis ex albo flavicantibus. *Briffon quad.* 107.

Le Petit-Gris. *De Buffon*, x. 116. *tab.* xxv. LEV. Mus.

S. with plain ears: hair of a dull grey color, mixed with black, and often tinged with dirty yellow: belly and infides of the legs white: tail long, bufhy, grey, and ftriped with black. Size of a half-grown rabbet.

Inhabits the woods of *North America*, *Peru* *, and *Chili* †; are very numerous in *North America*; do incredible damage to the plantations of *Mayz*; run up the ftalks, and eat the young ears; defcend in vaft flocks from the mountains, and join thofe that inhabit the lower parts; are profcribed by the provinces, and a reward of three pence *per* head for every one that is killed; fuch a

---

* *Chinchilles* are fmall beafts, like fquirrels, with wonderful fmoothe and foft fkins, which they weare as a healthfull thing to comfort the ftomacke ; they make coverings and rugs of the haire of thefe *Chinchilles*, which are found on the *Sierre* of *Peru. Acofta* in *Purchas's Pilg.* iii. 966.

† *Ovalle*, in his hiftory of *Chile*, fays, that the grey or afh-color'd fquirrels, of the valley of *Guafco*, are valuable for the furs. *Churchill's Coll.* vol. iii. 44.

number

1 *Hudsons Bay Squirrel* ___ N.° 341.

2. *Black* ___ N.° 339.   3. *Grey* ___ N.° 338.

number was deftroyed one year, that *Penfylvania* alone paid in rewards 8000*l.* of its currency.

Make their nefts in hollow trees, with mofs, ftraw, wool, &c. Feed on the mayz in the feafon, and on pine-cones, acorns, and mafts of all kinds. Form holes under ground, and there depofit a large ftock of winter provifion. Defcend from the trees and vifit their magazines when in want of meat; are particularly bufy at the approach of bad weather; during the cold feafon keep in their nefts for feveral days together; feldom leap from tree to tree, only run up and down the bodies; their hoards often deftroyed by fwine; when covered with deep fnow, the fquirrels often perifh for want of food; are not eafily fhot, nimbly changing their place, when they fee the gun levelled; have the actions of the common fquirrel; eafily tamed; their flefh efteemed very delicate. The furs which are imported under the name of *petitgris* are valuable, and ufed as linings to cloaks.

Quahtechalotl-thlitic. *Hernandez Mex.* 582. *Hernandez Nov. Hifp.* 8. Black Squirrel. *Catefby Car.* ii. 73.

L'Ecureuil noir. *Briffon quad.* 105. Sciurus niger. *Lin. fyft.* 86. Lev. Mus.

339. Black.

S. with plain ears: fometimes wholly black, but often marked with white on the nofe, the neck, or end of the tail: the tail fhorter than that of the former: the body equal.

Inhabits the *North of Afia, North America,* and *Mexico.* I fhould have placed it as a variety of the laft fpecies, did not Mr.

U   *Catefby*

Catefby exprefsly fay, that it breeds and affociates in feparate troops; is equally numerous with the former; commits as great ravages among the *Mayz*; makes its neft in the fame manner, and forms, like them, magazines for winter food.

A fquirrel of a moft beautiful fhining black color, is found at the *Pulo Condore*, in lat. 8.′ 40.

β. SQUIRREL, with plain ears : coarfe fur, mixed with dirty white, and black, but varies to white: throat, and infide of the legs and thighs, black : tail much fhorter than thofe of fquirrels ufually are: of a dull yellow color, mixed with black: body of the fize of the grey fquirrel. LEV. MUS.

Inhabits *Virginia*; defcribed from Mr. *Knaphan*'s collection; who told me that the planters called it the *Cat Squirrel*.

340. MADAGAS-
CAR.

S with plain ears: color of the face, back, fides, tail, and outfide of the limbs, of a dark glofsy black : ears, end of the nofe, cheeks, and all the under fide of the limbs, yellowifh white. The length of this fpecies from the tip of the nofe to the origin of the tail, is about eighteen inches: the tail is longer than the body, flender, and ends in a point.

Inhabits *Madagafcar*: defcribed by M. *de la Cepede*, in his fupplement to M. *de Buffon*, vii. 256. tab. lxxiii.

3

S. with

S with plain ears: fmaller than the *European:* marked along the middle of the back with a ferruginous line from head to tail: the fides paler: belly of a pale afh-color, mottled with black: tail not fo long, or fo full of hair, as the common kind; of a ferruginous color, barred with black, and towards the end is a broader band of the fame color. Lev. Mus.

Inhabits the pine-forefts about the *Bay,* and *Terra de Labrador.*

341. Hudson's Bay.

&. Carolina * Squirrel, with the head, back, and fides grey, white, and ruft-colored intermixed: belly white, divided from the colors of the fides by a ferruginous line: lower part of the legs red: tail brown, mixed with black, and edged with white.

Thefe are rather leffer than the *European* fquirrels: vary in the colors: in moft the grey predominates.

Quauhtecollotlquapachtli. *Hernandez Nov. Hifp.* 8.
Le Coquallin. *De Buffon,* xiii. 109. *tab.* xiii.

342. Varied.

S with plain ears: upper part of the body varied with black, white and brown: the belly tawny †: twice the fize of the common fquirrel.

* Leffer Grey Squirrel of the old edition.
† Called by the *Indians, Coztiocotequallin,* or Yellow Belly.

U 2   Inhabits

Inhabits *Mexico :* lives under ground, where it brings forth its
young, and lays in a ftock of winter food : lives on mayz : is
never to be tamed.

Thefe probably vary in fize : I have feen one that feemed to
be of this fpecies, but not fuperior in fize to the common fquirrel : the colors were brown, orange, and cinereous : the belly
orange.

343. FAIR. Sciurus flavus. Sc. auriculis fubrotundis, pedibus pentadactylis, corpore luteo. *Lin.*
fyft. 86. *Amœn. Acad.* i. 561.

S. with the body and tail of a flaxen color: of a very fmall
fize, with plain round ears, and rounded tail.

Inhabits the woods near *Amadabad,* the capital of *Guzarat,* in
great abundance, leaping from tree to tree *. *Linnæus* fays it is.
an inhabitant of *South America.*

344. BRASILIAN. Sciurus Brafilienfis ? *Marcgrave Brafil.*   107.
330.   Sciurus æftuans. Sc. grifeus, fubtus flá-
Sciurus coloris ex flavo et fufco mixti   vefcens. *Lin. fyft.* 88.
tæniis in lateribus albis. *Briffon quad.*

S. with plain ears, and rounded tail: head, body, and fides,
covered with foft dufky hairs, tipt with yellow : tail rounded: the hairs annulated with black and yellow : throat cinere-

* L'Ecureuil blond. *Della Valle,* p. 84.

ous :

ous: infide of the legs, and the belly, yellow: the belly divided lengthways with a white line; which begins on the breaft, is interrupted for a fmall fpace in the middle, and is then continued to the tail: length, from nofe to tail, eight inches one quarter: tail ten.

Inhabits *Brafil* and *Guiana.* Mr. *Vandeck,* captain of a man of war in the *Portuguefe* fervice, who procured them from their fet-tlements in *S. America,* favored me with two.

Tlalmototli. *Hernandez Nov. Hifp.* 9.   *Muf.* i. 76. *tab.* xlvii. fig. *2, 3. Briffon*   345. MEXICAN.
Sciurus rariffimus ex Nov. Hifpania. *Seb.*   *quad.* 103.

S of a moufe-color: the male marked on the back with
• feven white lines, which extend along the tail; the female, with only five: the tail of the male divided into four parts at the end: perhaps accidentally: its *fcrotum* pendulous, like a goat's.

Inhabits *New Spain. Seba,* in *tab.* xlviii. *fig.* 5. has the figure of another, of an uniform color, diftinguifhed alfo by its vaft *fcrotum.*

Muftela Africana. *Clus. Exot.* 112. *Raii*   Sc. palmarum. Sc. coloris ex rufo et ni-   346. PALM.
   fyn. quad. 216.   gro mixti, tæniis in dorfo flavicantibus.
Sciurus palmarum. Sc. fubgrifeus ftriis   *Briffon quad.* 109.
   tribus flavicantibus, caudaque albo ni-   Le Palmifte. *De Buffon,* x. 126. *tab.*
   groque lineata. *Lin. fyft.* 86.   xxvi.

S with plain ears: an obfcure pale yellow ftripe on the middle
• of the back, another on each fide, a third on each fide of the

the belly; the two laſt at times very faint: reſt of the hair on the ſides, back, and head, black and red, very cloſely mixed; that on the thighs and legs more red: belly, pale yellow: hair on the tail does not lie flat, but encircles it; is coarſe, and of a dirty yellow, barred with black. Authors deſcribe this kind with only three ſtripes: this had five, ſo poſſibly they vary.

Governor *Loten* did me the favor of informing me that it lived much in the *Coco* trees, and was very fond of the *ſury*, or palm-wine, which is procured from the tree; from which it obtained, among the *Indians*, the name of *Suricatsje*, or the *little cat* of the *Sury* *.

According to *Cluſius* and Mr. *Ray*, this ſpecies does not erect its tail like other ſquirrels, but has the faculty of expanding it ſideways.

847. WHITE-STRIPED.

β. BARBARY. Sciurus getulus. *Caii opuſc.* 77. *Geſner quad.* 847. Sc. getulus. Sc. fuſcus ſtriis quatuor albis longitudinalibus. *Lin. ſyſt.* 87. *Klein quad.* 84. *Briſſon quad.* 109. Barbarian ſquirrel. *Edw.* 198. Le Barbareſque. *De Buffon,* x. 126. *tab.* xxvii.

S. with full black eyes and white orbits: head, body, feet, and tail, cinereous, inclining to red: lighteſt on the legs: ſides, marked lengthways with two white ſtripes: belly white: tail buſhy, marked regularly with ſhades of black, one beneath the other: ſize of the common ſquirrel.

* See the proceſs of obtaining this liquor in *Rumphius's herbarium Amboinenſe,* vol. i. p. 5. The tree is engraved in *tab.* i. ii.

Both

Both thefe fquirrels inhabit *Barbary* and other hot countries :
live in trees, efpecially *palms,* from which one takes its name.

THIS fpecies refembles much the common fquirrel, but is
lighter colored, and has a yellow line extending along the
fides, from leg to leg.

Common in *Java* and *Prince*'s ifland; is called by the *Malayes,*
Ba-djing; lives much on *Plantanes*; is very fhy; retreats at the
fight of mankind, and clatters over the dry leaves of the *Pitang*
or *Plantanes* with vaft noife. It alfo is common on the *tamarind*
trees.

A. with membranes from fore leg to hind leg.

Sciurus Sagitta. Sc. hypochondriis prolixis volitans, cauda plano-pinnata lan-
ceolata. *Lin. fyft.* 88.
Sciurus petaurifta. *Pallas Mifcel. Zool.*
54. *tab.* vi.
Sciurus maximus volans, feu felis volans.
Sc. caftanei coloris, in parte corporis
fuperiore, in inferiore vero eximié fla-
vefcentis; cute ab anticis cruribus ad
poftica membranæ in modum extenfa
volans. *Briffon quad.* 112.
Le 'Taguan ou grand Ecureuil volant. *De*
Buffon, Suppl. iii. 150. tab. xxi. *Muf.*
Roy. Society *.

S with a fmall rounded head : cloven upper lip : fmall blunt
ears : two fmall warts at the outmoft corner of each eye,

* Where there is the fkin of one in fine prefervation.

with

with hairs growing out of them: neck fhort: four toes on the fore feet; and inftead of a thumb, a flender bone, two inches and a half long, lodged under the lateral membrane, ferving to ftretch it out: from thence to the hind legs extends the membrane, which is broad, and a continuation of the fkin of the fides and belly; the membrane extends along the fore legs, and ftretches out near the joint in a winged form: five toes on the hind feet, and on all the toes fharp, compreffed, bent claws: tail covered with long hairs, difpofed horizontally: color of the head, body, and tail, a bright bay; in fome parts inclining to orange: breaft and belly of a yellowifh white: length, from nofe to tail, eighteen inches; tail fifteen.

Inhabits *Java**, and others of the *Indian* iflands: leaps from tree to tree as if it flew: will catch hold of the boughs † with the tail. Differs in fize: that defcribed by *Linnæus* was the fize of our fquirrel: that killed by Sir *Edward Michelbourne*, in one of the *Indian* ifles, was greater than a hare. *Nieuhoff* defcribes this fpecies under the name of the Flying Cat, and fays the back is black: he has given two very good figures of it; one in his frontifpiece, the other in the page he defcribes it in ‡.

* *Hamilton's voy.* ii. 131.
† *Sir Edward Michelbourne's voy. in Purchas's Pilgrim.* i. 134.
‡ *Churchill's coll.* ii. 354.

Greater

Sailing Squirrel ____ N.º 349.

Greater Flying Squirrel. *Ph. Tr.* lxii. 379.

S. with back and fides of a deep cinereous color at the bottom; end ferruginous: under fide of the body of a yellowifh white; the hair every where long and full: tail covered
with long hairs, difpofed in a lefs flat way than thofe of the *European* kind; brown on the upper part, darker at the end, yellowifh beneath the fkin: the inftrument of flying difpofed from
leg to leg; but does not border the fore-legs.

Size equal to the *European* fquirrel.

SIZE.

Inhabits the fouthern parts of *Hudfon's Bay*, about *Severn* river.
Muf. Roy. Society.

PLACE.

Affapanick. *Smith's Virginia,* 27. *Joffelyn's voy.* 86. *De Laet,* 88.
Sciurus Americanus volans. *Raii fyn. quad.*
215. *Lawfon's Carolina,* 124. *Catefby
Carolina,* ii. 76, 77. *Edw.* 191. *Kalm,*
i. 321. *tab.* i. *Du Pratz.* ii. 69.
Sciurus volans. Sc. hypochondriis prolixis
volitans, cauda rotundata. *Lin. fyft.* 88.
Sciurus volans. *Briffon quad.* 110. iii. No.
12. LEV. MUS.

351. FLYING.

S. with round naked ears: full black eyes: a lateral membrane from fore to hind legs: the fore legs for the moft
part clear of the membrane: tail with long hairs difpofed horizontally, longeft in the middle, and ending in a point: color
above, a brownifh afh: beneath white, tinged with yellow. Much
lefs than the common fquirrel.

VOL. II.   X   Inhabits

SQUIRREL.

Inhabits *North America* and *New Spain* *: lives in hollow trees : fleeps in the day; during the night very lively; is gregarious, numbers being found in one tree: leaps from bough to bough fometimes at the diftance of ten yards: this action improperly called *flying*, for the animal cannot go in any other direction than forward; and even then cannot keep an even line, but finks confiderably before it can reach the place it aims at: fenfible of this, the fquirrel mounts the higher, in proportion to the diftance it wifhes to reach: when it would leap, it ftretches out the fore legs, and extending the membranes, becomes fpecifically lighter than it would otherwife be; and thus is enabled to fpring further than other fquirrels that have not this apparatus. When numbers leap at a time, they feem like leaves blown off by the wind. Their food the fame as the other *American* fquirrels: are eafily tamed: bring three or four young at a time.

352. NORFOLK-ISLE.

Stockdale's Bot. Bay, 151. *White,* 288.

S. with very fhort ears, almoft hid in the fur: color very much refembling that of the *American* grey fquirrel: a black line extends from the head along the middle of the back to the tail: the flying membrane black, edged with white: two thirds of the tail are of an elegant afh-color; the reft black: fize of the *American* grey fquirrel.

Inhabits *Norfolk ifle.*

* Where it is called *Quimichpatlan. Hernandez, Nov. Hifp.* 8.

In

In the ifle of *Pulo Condore* is a flying fquirrel ftriped with brown and white: poffibly a new fpecies.

Sciurus Virginianus volans. *Seb. Muf.* i. *tab.* xliv. *Briffon quad.* iii. Mus volans. *Lin. fyft.* 85.

353. Hooded.

S with the lateral membrane beginning at the chin and ears, and extending like the former from fore to hind leg: reddifh above; cinereous, tinged with yellow, beneath: ears large and oval.

Inhabits *Virginia*, according to *Seba*; who is the only author who has defcribed it. *Linnæus*'s fynonyms, from *Ray* and *Edwards*, erroneous.

Mus Ponticus vel Scythicus. *Gefner quad.* 743.
Sciurus Petaurifta volans. *Klein quad.* 54.
Flying fquirrel. *Ph. Tranf. abr.* ix. 76. tab. v.
Sciurus volans. *Faun. fuec.* No. 38. *Pallas, nov. fp. fafc.* i. 355.
Sc. volans Sc. hypochondriis prolixis volitans, cauda rotundata. *Lin. fyft.* 88.
Sciurus Sibiricus volans. *Briffon,* 11C. No. 13.
Le Poulatouche. *De Buffon,* x. 95. tab. xxii.
Quadrupes volatilis *Ruffiæ.* Com. acad. *Petrop.* v. 218. Lev. Mus.

354. European Fl. Sq.

S with naked ears, indented on the exterior fide: full eyes: eyelids bordered with black: membranes extend to the very bafe of the fore feet, and form a large wing on the exterior fide: tail full of hair, and round at the end: color of the

X 2

upper

upper part of the body a fine grey, like that on a gull's back: lower part of a pure white.

SIZE.

From nofe to tail four inches and a quarter; of the tail to the tip of the hair, five.

PLACE.

Inhabits *Finland* and *Lapland*, and the *Ruffian* dominions, from *Livonia* to the river *Kolyma* or *Kowyma*, in the N. E. part of *Siberia*, and is common in all the mountanous wooded tracts of that cold region: lives ufually on birch-tree buds and fructifications, and on the cones of the pines and cedars: is not gregarious, and leads a folitary life, and wanders about even in winter: lives in hollow trees, and makes its neft in the mofs of birch-trees: when at reft, it flings its tail over its back; but in leaping, extends it.

NAMES.

The *Germans* call it *Konige der Grauwerke**, or King of the Squirrels; the *Ruffians*, *Polatucha*, and *Letaga*; the *Poles*, *Wieiviorka Lataiaca*.

* *Klein.*

Two

Two cutting teeth in each jaw.
Four toes before: five behind.
Naked ears.
Long tail, covered with hair.

Moufe fquirrel. *Joffelyn's voy.* 86.
Ground fquirrel. *Lawfon Carolina,* 124.
Catefby Carolina, ii. 75. *Edw.* 181.
Kalm, i. 322. *tab.* i.
Sciurus *Lifteri. Raii fyn. quad.* 216.
Sciurus minor virgatus. *Nov. Com. Petrop.*
'v. 344.
Boern-doefkie. *Le Brun, voy. Mofcov.* ii.
342.

Sciurus ftriatus. Sc. flavus ftriis quinque
fufcis longitudinalibus. *Lin. fyft.* 87.
Klein quad. 53. *Pallas nov. fp fafc.* i.
373.
Sciurus Carolinenfis. *Briffon quad.*
Le Suiffe. *De Buffon,* x. 126. *tab.* xxviii.
Charlevoix Nouv. France, v. 198. LEV.
Mus.

355. STRIPED.

D. with plain ears: ridge of the back marked with a black
ftreak: each fide with a pale yellow ftripe, bounded
above and below with a line of black: head, body, and tail, of
a reddifh brown; the tail the darkeft: breaft and belly white:
nofe and feet pale red: eyes full.

Inhabits the north of *Afia*, beginning about the river *Kama*,
and grows more and more frequent in the woody parts of *Siberia*; but found in the greateft abundance in the forefts of *North
America*: they never run up trees except purfued, and find no
other means of efcaping: they burrow, and form their habitations
under ground with two entrances, that they may get accefs to the
one, in cafe the other is ftopped up. Their retreats are formed with
great fkill, in form of a long gallery, with branches on each fide,
each of which terminates in an enlarged chamber, as a magazine

PLACE.

MANNERS.

MAGAZINES.

4

to

to ftore their winter provifion in; in one they lodge the acorns, in another the *mayz*, in a third the hickery nuts, and in the laft, their favorite food, the *chinquapin* chefnut. They very feldom ftir out during winter, at left as long as their provifions laft; but if that fails, they will dig into cellars where apples are kept, or barns where *mayz* is ftored, and do a great deal of mifchief; but at that time the cat deftroys great numbers, and is as great an enemy to them as to mice.

During the *mayz* harveft, thefe animals are very bufy in biting off the ears, and filling their mouths fo full with the corn, that their cheeks are quite diftended, having pouches in their jaws like the *hamfter*. It is obfervable, that they give great pre-ference to certain food; for if, after filling their mouths with rye, they happen to meet with wheat, they fling away the firft, that they may indulge in the laft. They are very wild, bite fe-verely, and are fcarcely ever tamed: the fkins are of little ufe; but are fometimes brought over to line cloaks.

356. Fat.

Glis. *Gefner quad.* 550. *Raii fyn. quad.* 229.
Glis vulgaris. *Klein quad.* 56.
Glis fupra obfcurè cinereus, infra ex albo cinerefcente. *Briffon quad.* 113.

Sciurus Glis. Sc. canus fubtus albidus. *Lin. fyft.* 87.
Le Loir. *De Buffon,* viii. 158. *tab.* xxiv.
Mus Glis. *Pallas nov. fp. fafc.* i. 88.

D. with thin naked ears: body covered with foft afh-colored hair: belly whitifh: tail full of long hair: from nofe to tail, near fix inches; tail four and a half: thicker in the body than the fquirrel.

Inhabits *France* and the fouth of *Europe*. Is found in the woods in the fouth-weft parts of *Ruffia*, and was difcovered by

Doctor

Doctor *Pallas* in the rocky caverns about the rivers *Samara* and *Volga.* The late Doctor *Kramer* favored me with one from *Auftria.* Lives in trees, and leaps from bough to bough: feeds on fruits and acorns: lodges in the hollows of trees: remains in a torpid ftate during winter, and grows very fat,

> *Tota mihi dormitur hyems, et pinguior illo*
> *Tempore fum, quo me nil nifi fomnus alit*.*

Was efteemed a great delicacy by the *Romans,* who had their *Gliraria,* places conftructed to keep and feed them in. I think that the *Italians* at prefent eat them.

Mus avellanarum major. *Gefner quad.* 735. Greater Dormoufe, or Sleeper. *Raii fyn. quad.* 219.
Glis fupra obfcurè cinereus, infra ex albo cinerefcens, macula ad oculos nigra. *Briffon quad.* 114.

Mus quercinus. M. cauda elongata pilofa, macula nigra fub oculos. *Lin. fyft.* 84.
Le Lerot. *De Buffon,* viii. 181. *tab.* xxv.
Mus nitedula. *Pallas, nov. fp. fafc.* i. 88.

357. GARDEN.

D. with the eyes furrounded with a large fpot of black, reaching to the bafe of the ears, and another behind the ears : head and whole body of a tawny color: the throat and whole under fide of the body white, tinged with yellow : the tail long : the hairs at the beginning very fhort; at the end bufhy : length, from nofe to tail, not five inches : the tail four.

Inhabits *France* and the fouth of *Europe:* is found in magpies nefts and hollow trees about the *Volga,* and other temperate

---

* *Martial Epig. Lib.* xiii. *Ep.* 59.

and

and fouthern parts of the *Ruffian* dominions. Neither this nor the former fpecies extend beyond the *Uralian* mountains: infefts gardens, and is very deftructive to fruits of all kind: is particularly fond of peaches: lodges in holes in the walls: brings five or fix young at a time: like the former, remains torpid during winter: has a ftrong fmell, like a rat.

358. DEGUS.

Sciurus Degus. *Molina Chili,* 284.

D. of a dull white color, and with a blackifh line crofs the fhoulders, reaching to the elbows: the tail ending in a tuft: ears rounded: larger than the common rat.

MANNERS.

Inhabits *Chili,* and lives under ground, near the hedges and bufhes; and forms its retreat into various galleries communicating with each other: feeds on roots and fruits, and lays up a large provifion of them for winter food. Is not torpid during that feafon like our dormoufe.

359. COMMON.

Mus avellanarum minor, the Dormoufe or Sleeper. *Raii fyn. quad.* 220.
Rothe Wald Maufs. *Kramer Auftria,* 317.
Glis fupra rufus, infra albicans. *Briffon quad.*
Mus avellanarius. M. cauda elongata pilofa, corpore rufo, gula albicante, pollicibus pofticis muticis. *Lin. fyft.* 83. *Faun. fuec.* No. 35. *Pallas nov. fp. fafc.* i. 89.
Le Mufcardin. *De Buffon,* viii. 193. *tab.* xxvi.
Dormoufe. *Edw.* 266. *Br. Zool.* i. 95. LEV. MUS.

D. with round naked ears: full black eyes: body of a tawny red: throat white: fize of a moufe, but plumper: tail two inches

inches and a half long, and pretty hairy, efpecially towards the end.

Inhabits *Europe*: lives in thick hedges: makes its neft in the hollow of a low tree, or in a thick bufh near the bottom, of grafs, mofs, or dead leaves: brings three or four young at a time: feldom appears far from its retreat: forms magazines of nuts: eats its food fitting up, like a fquirrel: at approach of winter, retires and rolls itfelf up, lying torpid: fometimes in a warm day revives, takes a little food, and relapfes into its former ftate.

D. with a flat head, obtufe nofe, eyes full and black, upper lip bifid. **360. EARLESS.**

Auricles very minute, fcarcely apparent: long whifkers.

Head, back, fides, and front of the fore legs, pale ferruginous, mixed with black: from fhoulder to hind parts, on each fide, a white line: above each eye another: belly and feet of a dirty white.

Tail black in the middle; hoary on the fides.

Toes long and diftinct: the knob on the fore feet large: claws very long.

Hind legs black behind, and naked.

Size of a common fquirrel, but much broader and flatter. **SIZE.**

800 miles above the *Cape of Good Hope*, about the mountain **PLACE.**
Sneeburgh. Communicated by Sir *Jofeph Banks*.

VOL. II. Y Never

MANNERS.   Never climbs trees: burrows, feeds on bulbous roots, and is particularly fond of potatoes: walks often on its hind feet; and often lies flat on its belly: very tame, and never offers to bite: frequently flirts up its tail: makes a warm nest, and forms in it a round hole, in which it lodges, and pulls to the orifice a quantity of materials, in order to close it: keeps sometimes in its retreat for three entire days.

361. GILT-TAIL.   Le Lerot a queue doree. *Allamand Supplem.* iv. 164. tab. lxvii.

D. with short broad ears, great whiskers, the face marked lengthways with a gold color line extending from the

SIZE.   nose to the space between the ears. The rest of the head and whole body and beginning of the tail are a purplish chesnut color,

PLACE.   the remaining half of the tail is black: the rest of a beautiful gold color. The tail is thick about the base.

Length from nose to tail is five inches; of the tail six and nine lines.

Inhabits *Surinam.* Lives on fruits and climbs up the trees.

GUERLINGUETS.   *M. de la Cepede* * gives us the description of two species of animals, which he calls *Guerlinguets.* He denies that they are true squirrels: the ears are naked, and the tail grows taper, yet is covered with long hair, but by no means disposed like

* Supplem. &c vii. 261. tab. lxv. lxvi.

that

Gilt tail Dormouse _____ *N°.361.*

that on the tail of the fquirrel : they may come into this genus;
at leſt let them remain here till we are better informed.

The larger is between feven and eight inches long, excluſive
of the tail : the tail is of equal length : the hair on the body
is very ſhort, and at its extremity a bright bay. The tail is rayed
with brown and tawny.

362. Greater.

The leſſer is little more than four inches long : the tail little
more than three: the body, legs, and tail, are clouded with olive
and aſh-color : o the face, lower part of the belly, and ſides of
the legs are tawny.

363. Lesser.

Two

XXXII.
JERBOA.

Two cutting teeth in each jaw.

Two very fhort fore legs: two very long hind legs, refembling thofe of cloven-footed water-fowl.

Very long tail, tufted at the end.

364. ÆGYPTIAN.

Μυς δίπυς. *Theophr. opufc.* 295. *Ælian*
hift. an. lib. xv. *c.* 26.
Mus bipes. *Plinii lib.* x. *c.* 65. *Texeira's*
Travels, 21.
Gerbua. *Edw.* 219. *Plaifted's journal,* 59.
Mus jaculus. M. cauda elongata floccofa,
palmis fubpentadactylis, femoribus lon-
giffimis, brachiis breviffimis. *Lin. fyft.*
85. *Haffelquift itin.* 198.
Le Jerbo. *De Buffon,* xiii. 141.
Mus fagitta. *Pallas nov. fp. fafc.* i. 306.
tab. xxi.

J with thin, erect, and broad ears: full and dark eyes: long
whifkers: fore legs an inch long; five toes on each; the
inner, or thumb, fcarce apparent; but that, as well as the reft,
furnifhed with a fharp claw: hind legs two inches and a quarter
long, thin, covered with fhort hair, and exactly refembling
thofe of a bird; three toes on each, covered above and below
with hair; the middle toe the longeft; on each a pretty long
fharp claw: length, from nofe to tail, feven inches and one quarter: tail ten inches, terminated with a thick black tuft of hair;
the tip white; the reft of the tail covered with very fhort coarfe
hair: the upper part of the body thin, or compreffed fideways:
the part about the rump and loins large: the head, back, fides,
and thighs, covered with long hair, afh-colored at the bottom,
pale tawny at the ends: breaft and belly whitifh: acrofs the upper

per part of the thighs is an obfcure dufky band : the hair long and foft.

Inhabits *Ægypt, Barbary, Paleftine,* the deferts between *Baf-fora* and *Aleppo,* the fandy tracts between the *Don* and *Volga,* the hills fouth of the *Irtifh,* from fort *Janiyfchera* to the *feven palaces,* where the *Altaic* mountains begin : as fingular in its motions as in its form : always ftands on its hind feet ; the fore feet performing the office of hands : runs faft ; and when purfued, jumps five or fix feet from the ground : burrows like rabbets : keeps clofe in the day : fleeps rolled up : lively during night : when taken, emits a plaintive feeble note : feeds on vegetables : has great ftrength in its fore feet. Two, which I faw living in *London,* burrowed almoft through the brick wall of the room they were in ; came out of their hole at night for food, and when caught, were much fatter and fleeker than when confined to their box.

This is the *Daman Ifrael,* or the *Lamb of the Ifraelites* of the *Arabs,* and is fuppofed to be the *Saphan* *, the coney of HOLY WRIT : our rabbet being unknown in the *Holy Land.* Dr. *Shaw* met with this fpecies on mount *Libanus,* and diftinguifhes it from the next fpecies †. It is alfo the moufe of *Ifaiah* ‡, *Achbar* in the original fignifying a male *Jerboa.*

This and the following fpecies, which is found to extend to the

---

* *Bochart* difplays a vaft deal of learning on the fubject. *Vide Hierozoicon, lib.* iii. *c.* 33. *p.* 1001.

† *Travels,* 376.

‡ Chap. lxvi. 17. *Bochart,* 1015. This animal was a forbidden food with the *Ifraelites.*

colder

colder regions, on any approach of cold grow torpid, and remain fo till they are revived by a change of weather. *Pallas* calls this clafs the *Species Lethargicæ.*

365. SIBERIAN.

Cuniculus pumilio faliens cauda longif-fima. *Nov. Com. Petrop.* v. 351. *tab.* ix. *fig.* 1.
Cuniculus pumilio faliens, cauda anoma-la lorgifſima. *Briſſon quad.* 103.

Dipus Jaculus. *Gm. Lin.* 157.
Flying hare. *Strahlenberg's hiſt. Ruſſ.* 370.
Mus jaculus. *Pallas nov. ſp. faſc.* i. 275. tab. xx. MUS. LEV.

α. GREAT.

NOSE truncated; end edged with white: lower teeth flender; twice as long as the upper.

Ears large, pointed, tipt with white, naked within: hairs on the back tawny, of a dark grey beneath, very foft: legs and whole under fide of the body white: half the tail next to the body covered with fhort whitifh hairs; from thence, with long black hairs; the end has a large white feathered tuft an inch long.

Five toes on the fore feet; the toe without a nail.

On the hind legs, an inch above the feet, are two long toes armed with nails: the back part of the legs naked.

Length eight inches and a half; of the tail ten.

PLACE.

This variety is no where very frequent, but is chiefly found from the *Caſpian* fea to the river *Irtiſh.*

β. MIDDLE.

Of the fize of a rat: of the color of the former, except that the rump on each fide is croffed with a white line.

This middle fpecies is found only in the eaftern deferts of *Siberia* and *Tartary*, beyond lake *Baikal*; alfo in *Barbary**** and *Syria*†, and even as far as *India* ‡.

* *Shaw's Travels.*   † *Haym's Teſoro Brit.* ii. p. and tab. 124.   ‡ *Pallas.*

Differs

Siberian Jerboa — N.365.

Differs from the *Great*, in wanting the white circle round the nose, in having a lefs tuft to the tail, and the end juft tipt with white: agrees entirely in form; but is far inferior in fize to even the *Middle*. Inhabits the fame places with the *Great*.

Thefe three agree in manners: burrow in hard ground, clay or indurated mud: not only in high and dry fpots, but even in low and falt places. They dig their holes very fpeedily, not only with their fore feet but with their teeth, and fling the earth back, with their hind feet, fo as to form a heap at the entrance. The burrows are many yards long, and run obliquely and winding, but not above half a yard deep below the furface. They end in a large fpace or neft, the receptacle of the pureft herbs. They have ufually but one entrance; yet by a wonderful fagacity they work from their neft another paffage to within a very fmall fpace of the furface, which in cafe of neceffity they can burft through, and fo efcape.

It is fingular, that an animal of a very chilly nature, fhould keep within its hole the whole day, and wander about only in the night.

They fleep rolled up, with their head between their thighs: and when kept in a ftove, and taken fuddenly out, they feem quite ftupified, and for a time fcarcely find the ufe of their limbs: perhaps this arifes from an excefs of heat; for when an attempt is made to take them out of their burrows, they are quickly alarmed on the noife of digging, and attempt their efcape. At fun-fet they come out of their holes, clear them of the filth, and keep abroad till the fun has drawn up the dews from the earth. On approach of any danger, they immediately take to flight, with leaps a fathom in height, and fo fwiftly that a

man

man well mounted can hardly overtake them. They fpring fo nimbly, that it is impoffible to fee their feet touch the ground. They do not go ftrait forwards, but turn here and there, till they gain a burrow, whether it is their own, or that of another. In leaping, they carry their tails ftretched out: in ftanding, or going or walking, they carry them in form of an S, the lower part touching the ground, fo that it feems a director in their motions. When furprized, they will fometimes go on all fours, but foon recover their attitude of ftanding on their hind legs like a bird: even when undifturbed, they ufe the former attitude; then rife erect, liften, and hop about like a crow. In digging or eating they drop on their fore legs: but in the laft action will often fit up and eat like a fquirrel.

EASILY TAMED.

They are eafily made tame: feek always a warm corner: fore-tell cold or bad weather by wrapping themfelves clofe up in hay; and thofe which are at liberty ftop up the mouths of their burrows.

FOOD.

In a wild ftate they are particularly fond of the roots of tulips: live much on oleraceous plants: the fmall ftature of the *pygmy* kind is attributed to their feeding on faline plants. Thofe of the middle fize, which live beyond the lake *Baikal*, live on the bulbs of the *Lilium Pomponium*, and they gnaw the twigs of the *Robinia Carugana*. When confined, they will not refufe raw meat, and the entrails of fowls.

They are the prey of all leffer rapacious beafts. The *Arabs*, who are forbidden all other kinds of mice, efteem thefe the greateft delicacies: as thofe people often are difappointed in dig-ging after them, they have this proverb, " To buy a hole inftead " of a *Jerboa*."

4   The

The *Mongols* have a notion that they fuck the fheep : certain it is, they are during night very frequent among the flocks, which they difturb by their leaps.

The *Mongols* call this animal *Alagh-Daagha*. *Alagh* fignifies variegated, *Daagha,* a foal. The *Calmucs* call it *Jalma:* the great fort they ftyle *Morin Jalma,* or the *Horfe Jerboa;* the leffer fort, *Choïn Jalma,* or the *Sheep*.

NAMES.

They breed often in the fummer ; in the fouthern parts, in the beginning of *May:* beyond *Baikal,* not till *June.* They bring perhaps eight at a time, as they have fo many teats. They fleep the whole winter without nutriment. About *Aftracan,* they will fometimes appear in a warm day in *February:* but return to their holes on the return of cold.

Animals of this genus were certainly the *two-footed mice,* and the *Ægyptian mice,* of the ancients, which were faid to walk on their hind legs ; and ufe the fore inftead of hands. Thefe, with the plant *Silphium,* were ufed to denote the country of *Cyrene,* where both were found, as appears from the figures on a beautiful gold coin preferved by Mr. *Haym *,* and which I have caufed to be copied above the animal, in the plate.

Dipus fagitta. *Gm. Lin.* 158. *Pallas nov. fp.* 87, 206. tab. xxi. *Edw.* tab. 219.

366. ARROW.

J. with ears fhorter and broader than the preceding : nofe longer and lefs obtufe : toes before, three behind : coat thicker

* *Teforo Brit.* ii. 124.

and longer; a white band from the bafe of the tail to the junction of the thighs with the body: length from the tip of the nofe to the rump, little more than five inches; of the tail fix.

Inhabits *Barbary*, and all the north of *Africa, Ægypt, Arabia*, and *Syria*; and lives in the fandy deferts.

367. CAPE. Grand Gerbo. *Allamand de Buffon*, xv. Dipus Cafer. *Gm. Lin.* 159. *Miller's* plates.
118. *Journal Hiflorique*, 59. tab. xxxi.

J. with a fhort head: broad between the ears: mouth placed far below the upper jaw: lower very fhort: two great teeth in each: ears one-third fhorter than thofe of the common rabbet, thin and tranfparent: eyes large: whifkers great.

Fore legs fhort, five toes on each, with a great protuberance next to the inner toe: claws of the fore toes crooked, and two-thirds longer than the toes themfelves: claws of the hind toes fhort.

Color above tawny; cinereous below, mixed with long hairs pointed with black: two-thirds of the tail tawny, the reft black.

Length from nofe to tail one foot two inches; of the tail near fifteen inches; the ears near three.

Inhabits the great mountains far north of the *Cape of Good Hope*. It is called by the *Hottentots, Aerdmannetje*; and by the *Dutch, Springen Haas*, or the *Jumping Hare*.

It is very ftrong; will leap twenty or thirty feet at a time: its voice a grunting: when it eats, fits upright, with the legs extended horizontally, and with a bent back: ufes its fore feet to bring the food to its mouth; burrows with them, which it does

fo

fo expeditioufly as quickly to bury itſelf. `In ſleeping, it ſits with its knees ſeparate, puts its head between its hind legs, and with the fore legs holds its ears over its eyes.

Mus longipes. M. cauda elongata veſti- | Mus cauda longa veſtita, pedibus poſticis | 368. Torrid.
ta, palmis tetradactylis, plantis penta- | longitudine corporis, flavis. Muſ. Ad.
dactylis, femoribus longiſſimis. Lin. | Fr. 9.
ſyſt. 84.

J. with naked oval ears: long whiſkers: four toes on the fore feet: the hind feet the length of the body, thick, ſtrong and thinly haired: five toes on each foot: ſcarcely any neck: tail the length of the body, with very little hair on it: color of the up-per part of the body yellow; the lower white: ſize of a common mouſe.

Inhabits, according to *Linnæus*, the torrid zone*: mentioned by no other writer.

* *Habitat in torridis regionibus.*

Two

XXXIII. RAT.   Two cutting teeth in each jaw.

Four toes before; five behind.

Very flender taper tail; naked, or very flightly haired.

* Jerboid.

369. CANADA.   R. with the upper jaw projecting far beyond the lower: upper cutting teeth deeply divided by a longitudinal furrow: ears fmall, and hid in the fur, and placed far back: the three middle hind toes very long; thofe on each fide very fhort: color of the fur on the upper part of the head and body, light ruft; towards the bottom of the fur iron grey: belly whitifh: length from nofe to the tip of the tail fcarcely two inches: tail very flender; three inches and a half long.

MANNERS.   This animal inhabits the woods of *Canada.* Its hind legs have more of the *Jerboas,* than any of the reft of this genus; are very long: it goes upright on thefe, like the Jerboa; and its pace is leaping like that animal: is exceeding nimble, and is with difficulty caught, except it can be forced into the open grounds: is torpid during winter: wraps itfelf up like the dormoufe, and coils up its long tail; previoufly rolling itfelf into a round ball of clay, which it forms for its winter retreat.

I am indebted to Col. *Davies,* of the artillery, for the fight and the account of this curious animal.

R. with

R with a blunt nose: mouth placed far beneath: upper lip bifid: ears large, naked, rounded; fore legs short, furnished with four toes, and a tubercle instead of a thumb: hind legs long and naked, like some of the *Jerboas:* thumb short: toes long, slender, and distinct; the exterior the shortest.

The whole length of the animal eight inches: of which the tail is four and three quarters.

Color deep brown above, white beneath, separated the whole length on each side by a yellow line.

Inhabits *Hudson's Bay,* and the *Labradore* coast. Sent by Mr. Graham, and deposited in the Museum of the *Royal Society.*

Since I wrote the above, I find that Doctor *Pallas* has described this species under the title of *Mus Longipes* *. It inhabits also the sandy desert of *Naryn,* or *Ryn Pesky,* between the *Volga* and the *Yaik,* near the *Caspian Sea,* in *lat.* 46½. In this tract scarcely any thing grows except the *Torlok,* or *Pterococcus Aphyllus,* and a few other poor plants on which it feeds. Two were then taken sporting in the mid-day sun; they were both males, and attempted to escape to different holes. The burrows had three entrances running obliquely, and were about a yard deep; lined or plaistered with mud. In the bottom was neither nest nor provision of grafs.

The *Asiatic* animal differed in color from the *American,* being above of a light grey mixed with tawny, white below: these colors divided lengthways by a stripe of dusky red. The tail

---

* *Nov. sp. fasc.* i. 314. *tab.* xviii. B. Mus mindianus? *Itin.* ii. 702.

5   covered

covered with longer and loofer hair at the end than in the other parts: the foles of the feet clad with hair. This I could not well obferve in the fpecimen from *Hudfon's Bay*, as it was preferved in fpirits. *Linnæus* defcribes this fpecies under the title of *Mus longipes, Syft. nat.* 84. Doctor *Pallas*, with great reafon, fuppofes it to be the fame with the *Jird* of Doctor SHAW, which our learned countryman defcribes with the *Jerboa*. It agrees in colors with the above; in its long tail being better cloathed than that of a rat; and in its burrowing under ground. This is frequent in *Barbary*, and is reckoned there a good food *.

371. A. CIRCAS-SIAN.

To this I join, on the judgment of Doctor *Pallas*, another animal, which I defcribed at Nº 205 of the *Synopfis* of *Quadrupeds*, under the title of *Circaffian Marmot*, or.

M. with ears like thofe of mice: red fparkling eyes: fharp teeth: body long, and of an equal thicknefs: chefnut-colored hair, long, efpecially on the back: has fharp claws: tail long and bufhy: fore feet fhorter than the hind feet: fize of the *Hamfter*, Nº 324.

Inhabits the neighborhood of the river *Terek*, which flows out of *Circaffia* and falls into the *Cafpian Sea*: runs faft up hill, very flowly down: burrows, and lives under ground †.

* *Shaw's travels*, 248.
† *Schober's* memorab. Afiat. Ruffiæ in *Muller's* Samlung Ruff. viii. 124.

Mus

1. *Middle Jerboa* _____ *p. 166.*

2. *Tamarisk Rat* _____ *N.º 372.*

Mus Tamaricinus. *Pallas, nov. fp.* i. 322. *tab.* xix. *Itin.* ii. 702.

R. with an oblong head: great whifkers: nofe blunt: noftrils covered with a flap: teeth yellow: eyes large and brown: ears large, naked, and oval: neck fhort: fpace round the nofe and eyes, and beyond the ears, white: fides of the head and neck hoary: back and fides of a yellowifh grey: tips of the hairs brown: breaft and belly white: tail cinereous; above half annulated with brown: hind legs long: on the fore feet a warty tubercle inftead of a thumb.

Length to the tail above fix inches: tail not quite fo long.

Inhabits the lower falt-marfhes about *Saritfchikofka*, on the *Lower Yaik* or *Ural*, where they burrow beneath the knotty roots of the tamarifk bufhes; each burrow has two entrances, and is very deep: they feed only at night: out of numbers which were taken in traps placed before their holes, not a female was taken. Their food is the fucculent maritime tribe of plants, fuch as *Nitraria, Salfola,* and others, with which thofe deferts abound.

To this divifion of Rats I give the title of *Jerboid,* from the affinity it has to that genus in the length of the hind legs. To the other, *Murine,* as comprehending all the common fpecies of Rats and Mice.

SIZE.
PLACE.

** Murine.

** Murine.

373. BLACK.

Mus domeſticus major, quem vulgò Rattum vocant. *Geſner quad.* 731. *Raii ſyn. quad.* 217.
Mus Rattus, Mus Ciſtrinarius. *Klein quad.* 57.
Ratze. *Kramer Auſtr.* 316.
Mus cauda longiſſima obſcurè cinerea. *Briſſon quad.* 118.

Mus Rattus. M. cauda elongata ſubnuda, palmis tetradaƈtylis cum unguiculo pollicari, plantis pentadaƈtvlis. *Lin. ſyſt.* Ratta. *Faun. ſuec.* N° 33 *Br. Zool.* i. N° 27.
Le Rat. *De Buffon,* vii. 278. *tab.* xxxvi. *Pallas nov. ſp. faſc.* i. 93. LEV. MUS.

R. of a deep iron-grey color, nearly black: belly cinereous: legs duſky, almoſt naked: a claw, in the place of a fifth toe, on the fore feet: length, from noſe to tail, ſeven inches; tail near eight.

PLACE.   Inhabits moſt parts of *Europe:* of late, the numbers much leſſened, and in many places extirpated, by the next ſpecies: very deſtruƈtive to corn, furniture, young poultry, rabbets, and pigeons: will gnaw the extremities of infants when aſleep: breeds often in a year: brings ſix or ſeven young at a time: makes its neſt, in a hole near a chimney, of wool, bits of cloth, or ſtraw: will deſtroy and devour one another: its greateſt enemy is the weeſel. Firſt introduced by the *Europeans* into *South America*,* about the year 1544, in the time of the Viceroy *Blaſco Nunnez.* Is now the peſt of all that continent.

The word *Rattus* is modern. The *Romans* probably comprehended all kinds under the word *Mus.* The *Welſh* call this *Llygoden Frengig,* or the *French Mouſe,* which evinces it not to be a native

* *Garcilaſſo de la Vega,* 384. *Ovalle. Churchill's coll.* iii. 43.

of

of our ifland. There is a very minute variety of this kind about the *Volga*, in the deferts of the lower part (for they have not reached the upper) which fcarcely weighs feven drachms.

I cannot trace the original place of the black rat: none are found in *Siberia* or *Kamtfchatka*.

Rats (I know not of what fpecies) are found in the *Papuas* iflands, off *New Guinea**; but according to the account given by Doctor *Forfter*†, the common black rats fwarm in *Otaheite*, and other of the *Society* iflands, and are alfo met with in the other groupes of iflands, in *New Zealand*, and in *New Holland*. They feed in *Otaheite* on the fruits of the country, and are fo bold as even to attack the inhabitants when they are afleep. The natives hold them in the utmoft deteftation, as unclean animals, and will even avoid killing them, leaft they fhould be polluted by the touch. They will not even eat the bread-fruit thefe animals fhould happen to run over.

SOUTH-SEA ISLANDS.

Le Coypu. *Molina Chili.* 268. Mus Coypus. *Gm. Lin.* 125.

374. COYPU.

R. with round ears: nofe elongated, covered with whifkers: legs fhort: tail thick, and of a moderate length, well covered with hairs: two very fharp cutting teeth in each jaw: on the fore feet are five toes, all feparated; on the hind feet five, palmated: has the appearance of the otter in hair and fize.

This animal lives equally well in the water as on the land; and

MANNERS.

* Captain *Forreft*.　　† *Obfervations, &c.* 185, 187.

frequents alfo houfes: is eafily tamed, and very content in the domeftic ftate: attaches itfelf to thofe who treat it kindly: has a piercing cry on being abufed: the female brings forth five or fix young, which always follow her.

375. BROWN.

Mus cauda longiffima, fupra dilutè fulvus, infra albicans. Le Rat de Bois. *Briffon quad.* 120.
Le Surmulot. *De Buffon*, viii. 206. *tab.*

xxvii.
Norway Rat. *Br. Zool.* i. N° 26.
Mus Decumanus. *Pallas nov. fp. fafc.* i. 91. LEV. MUS.

R. with the head, back, and fides, of a light brown color, mixed with tawny and afh-color: breaft and belly dirty white: feet naked, and of a dirty flefh-color: fore feet furnifhed with four toes, and a claw inftead of the fifth: length, from nofe to tail, nine inches; tail the fame: weight eleven ounces: is ftronger made than the laft.

IN EUROPE.

Inhabits moft parts of *Europe:* but was a ftranger to that continent 'till the prefent century: came into *Great Britain* about fifty years ago: not known in the neighborhood of *Paris* half that time. This rat is common in *India,* both on the land and in fhips. May we not go to *Indoftan* for their origin? They fwarm in *Peterfburg:* have reached *Pruffia,* but not the oppofite fide of the *Baltic;* at left *Linnæus* takes no notice of them.

ASIA.

Are numerous in *Perfia,* where they burrow in the fields *. In *Hyrcania* they occupy the deferted holes of the porcupine.

---

* Doctor *Pallas,* among his other epiftolary communications.

Some

Some years ago an immenfe migration arrived from the weft at the town of *Jaik*; and in the year 1727 an equal number appeared about *Aftracan*, filled the whole bed of the *Volga*, and infefted the houfes to that degree, that nothing could be preferved from them. They have not yet reached *Siberia*. Thefe probably were the *Mures Cafpii* of *Ælian*, which he fays were little lefs than *Ichneumons*; and made periodical vifits in infinite multitudes to the countries bordering on the *Cafpian Sea:* fwimming boldly over the rivers, holding by one another's tail *.

Burrow, like the water-rat, on the fides of ponds and ditches: fwim well and dive readily: live on grain and fruits, and will deftroy poultry and game: encreafe faft; bring from fourteen to eighteen young at a time: are very bold and fierce; will turn when clofely purfued, and faften on the ftick or hand of thofe who offer to ftrike them: have deftroyed the common black rat in moft places. Inhabit fields part of the year, but migrate in great numbers into houfes, and do infinite mifchief.

Le Rat Perchal, *De Buffon, Supplem.* vii. 276. tab. lxix.

376. PERCHAL.

R with ears rounded on the top: nofe long and turning up: body longer than that of common rats: hair on the upper parts deep brown: hind legs larger than the fore: tail naked and fcaly: length from nofe to tail above a foot; tail between eight and nine inches.

Common in *India*, and infefts the houfes in *Pondicherry*,

PLACE.

* *Æliani hift. ar.* xvii. *c.* 17.

A a 2

as

as our rats do thofe of *Europe*: are very numerous: the inhabitants ufe them for food.

377. BANDICOTE. At p. 440 of the former edition I imagined that the *Brown rat* was the fame as the *Bandicote* of the *Eaft Indies*. My good and intelligent friend Doctor *Patrick Ruffel,* who has made a long refidence on the eaftern coaft of *Indoftan,* convinces me of my miftake. His remarks are fuch that do not at prefent enable me to give fo full a defcription of this fpecies as I could wifh. It is generally agreed that the *Bandicote* is at left five times the weight of the *Brown rat:* that, comparative with that kind, it has a fhorter and thicker tail: that its general form is much thicker, and the back arched; fo that, at firft fight, it looks like a little pig: it is lefs active and alert than the brown-rat: is infinitely mifchievous in gardens: burrows under the houfes, and will even undermine them fo as to caufe them to fall: never go on board fhips. The *Palinquin-boys* eat this kind, but will reject the common rat. A more fatisfactory account of the *Bandicote* may be expected in the courfe of a year.

378. AMERICAN. LEVERIAN MUSEUM.
Mus Caraco? *Pallas nov. fp. fafc.* i. 335. *tab.* xxiii.

R. with the upper jaw much longer than the lower: head long: nofe narrow and pointed: ears large and naked: whifkers fine, but long: tail naked, and like that of the black rat, but not fo long.

Color

Color a deep brown; on the belly inclines to afh-color: hair ruder than in the preceding fpecies.

In fize larger than the black, leffer than the brown rat.

Inhabits *North America*; but I am uncertain whether it is entirely wild, or whether it has yet found its way into houfes and out-houfes. Mr. *Bartram* * mentions the rat (but does not determine the fpecies) which lives among the ftones and caverns in the *Blue Mountains*, far from mankind: comes out at night, and makes a terrible noife; but in very fevere weather keeps filent within its holes.

β. CARACO.

The Mus CARACO of Doctor *Pallas* is fo nearly allied to this fpecies, that I do not at this time venture to feparate them : the whifkers of the former feem rather fhorter, and the tail, in proportion to its length, thicker; but the thinnefs of that part might, in the fpecimen in the LEVERIAN MUSUEM, arife from its being dried ; neither could I examine it thoroughly, as it was within a glafs cafe. The *Caraco* has not as yet appeared to the weft of the *Jenefei*, but fwarms about and beyond lake *Baikal*. It has much agreement with the laft kind, being, as the *Mongals* report, converfant among lakes and waters, and is called by them *Characho*, and *Jike-Cholgonach* or the *Great Moufe*. It burrows in the banks of rivers: is fuppofed to extend to *China*, and to be very noxious there.

* In *Kalm's trav.* ii. 48.

Le

379. SCHERMAN.   Le Scherman. *De Buffon, Supplem.* vii. 278. tab. lxx.

R. with a ſhort head and thick noſe: ſmall eyes: ears ſo very ſmall as to be ſcarcely viſible: color of the hair duſky, mixed with grey and tawny: edges of the mouth bordered with white: body ſix inches long; tail above two.

Common about *Straſbourg*, in the gardens and places near the water: make great havoc among the plants and the cultivated grounds: ſwim and dive very well, and are often taken by the fiſhermen in their weels: burrow under ground, and are frequently caught in the traps uſed by the people who are employed in taking the *Hamſter* rat.

380. WATER. Le Rat d'Eau. *Belon, Aquat.* 30. *tab.* xxxi.
Mus aquatilis. *Agricola An. Subter.* 488. *Geſner quad.* 732. *Raii ſyn. quad.* 217. *Klein quad.* 57.
Waſſer-maus. *Kramer Auſtr.* 316.
Mus Amphibius. M. cauda elongata piloſa, plantis palmatis. *Lin. ſyſt.* 82. *Faun. ſuec.* N° 32. *Pallas Nov. ſp. faſc.* i. 20.

M. cauda longa pilis ſupra ex nigro et flaveſcente mixtis, infra cinereis veſtitus. *Briſſon quad.* 1:4.
Le Rat d'Eau. *De Buffon,* vii. 348. *tab.* xliii.
Water Rat. *Br. Zool.* i. N° 27. LÉV. MUS.

R. with a thick blunt noſe: ears hid in the fur: eyes ſmall: teeth yellow: on each foot five toes; inner toe of the fore foot very ſmall; the firſt joint very flexible: head and body covered with long hairs, black mixed with a few ferruginous hairs: belly of an iron grey: tail covered with ſhort black hairs; the tip whitiſh: weight nine ounces: length, from noſe to tail, ſeven inches;

inches; tail only five: fhape of the head and body more compact than the former fpecies*.

Inhabits *Europe*, the north of *Afia*, and *North America*†; burrows in the banks of rivers, ponds, and wet ditches: feeds on fmall fifh, and the fry of greater; on frogs, infects, and roots: is itfelf the prey of pike: fwims and dives admirably, though it is not web-footed, as Mr. *Ray* fuppofed, and *Linnæus* copied after him: brings fix young at a time. This animal and the Otter are eaten in *France* on *maigre* days.

Le Guanque. *Molina,* 281. Mus Cyanus. *Gmelin,* 132.   381. SKY-COLORED.

R with rounded ears: fur of a blue color: fize and appearance of my field rat.

Inhabits *Chili*: burrows a gallery ten feet long, with feven correfpondent chambers on each fide of a foot in depth: thefe are the magazines for winter provifion, which are of roots, moft nicely laid in order one upon the other: at the approach of the rainy feafon retire to the burrows: breed twice in the year, and bring forth fix

---

* It has fome refemblance to the Beaver, which induced *Linnæus*, in the firft edition of his *Fauna Suecica,* to ftyle it *Caftor cauda lineari tereti.*

† *Lawfon hift. Carolina,* 122. He alfo mentions another, which he calls the *Marfh* Rat, being more hairy than the common rat; but apparently is the fame with this. Thofe of *Canada* vary to tawny and white. Vide *De Buffon,* xiv. 401. xv. 146.

at

at a time: the firft brood is left to provide for itfelf; the fecond re-tires under ground with the parents: are very timid, and very cleanly in their retreats: the peafants hunt for the hoards, and by robbing them leave the family to perifh.

382. Mouse.

Mus domefticus communis feu minor. *Gefner quad.* 714. *Raii fyn. quad.* 218.
Mus minor, mufculus vulgaris. *Klein quad.*
Maufs. *Kramer Auftr.* 316.
Mus mufculus. M. cauda elongata, palmis tetradactylis, plantis pentadactylis.

Lin. fyft. 83. *Pallas Nov. fp. fafc.* i. 95. Mus. *Faun. fuec.* N° 34.
Mus cauda longiffima, obfcurè cinereus, ventre fubalbefcente. *Briffon quad.* 119.
La Souris. *De Buffon,* vii. 309. *tab.* lix. *Br. Zool.* i. N° 30. Lev. Mus.

AN animal that needs no defcription: when found white, is very beautiful, the full bright eye appearing to great advantage amidft the fnowy fur.

Inhabits all parts of the world, except the *Arctic:* follows mankind.

383. Field.

Mus agreftis minor. *Gefner quad.* 733.
Mus domefticus medius. *Raii fyn. quad.* 218.
Maufs mit weiffen bauch. *Kramer Auftr.* 317.
Mus cauda longa fupra e fufco flavefcens, infra ex albo cinerefcens. *Briffon quad.* 123.

Mus fylvaticus. M. cauda longa, palmis tetradactylis, plantis pentadactylis, corpore grifeo pilis nigris abdomine albo. *Lin. fyft.* 84. *Pallas Nov. fp. fafc.* i. 94. *Faun. Suec.* N°36.
Le Mulot. *De Buffon,* vii. 325. *tab.* xli.
Long-tailed Field-moufe. *Br. Zool.* i. N°28. Lev. Mus.

R with full and black eyes: head, back, and fides of a yellowifh brown, mixed with fome dufky hairs: breaft of an ochre-color: belly white: length, from the tip of the nofe to the tail,

5   four

four inches and a half: tail four inches, flightly covered with hair.

Inhabits *Europe:* found only in fields and gardens: feeds on nuts, acorns, and corn: forms great magazines of winter provifion: hogs, tempted by the fmell, do much damage in the fields by rooting up the hoards: makes a neft for its young very near the furface, and often in a thick tuft of grafs: brings from feven to ten at a time: called, in fome parts of *England,* *Bean Moufe,* from the havock it makes among the beans when juft fown.

Is common in *Ruffia,* and about the *Urallian* chain, but not beyond.

æ. AMERICAN R. with very long whifkers, fome white, others black: ears large, naked, and open: from the head to the tail, along the middle of the back, a broad dark ftripe, ferruginous and dufky: the cheeks, fpace beneath the ears, and fides, quite to the tail, oránge-colored: under fide, from nofe to tail, of a fnowy whitenefs: feet white: hind legs longer than thofe of the *European* kind: tail dufky above, whitifh beneath. *New York.*

The lefs long-tailed Field-Moufe. *Br. Zool.* ii. *App.* 498. Lev. Mus.   384. HARVEST.

R. with eyes lefs prominent than thofe of the former: ears prominent: of a full ferruginous color ábove, white beneath: a ftrait line along the fides divides the colors: tail a little hairy: length, from nofe to tail, two inches and a half: tail two inches: weight one-fixth of an ounce.

Inhabits *Hampſhire*; where it appears in greateſt numbers during harveſt: never enters houſes; but is carried into the ricks of
corn in the ſheaves; and often hundreds are killed on breaking up
the ricks: during winter, ſhelters itſelf under ground: burrows
very deep, and forms a warm bed of dead graſs: makes its neſt
for its young above ground, between the ſtraws of ſtanding corn;
it is of a round ſhape, and compoſed of blades of corn: brings,
about eight young at a time.

385. ORIENTAL.

Mus orientalis. *Seb. Muſ.* ii. 22. *tab.* xxi.
fig. 2.
M. cauda mediocri ſubnuda, palmis tetradactylis, plantis pentadactylis, corporis ſtriis punctatis. *Lin. ſyſt.* 84.
M. cauda longa, ſtriis corporis longitudinalibus & punctis albis. *Muſ. Ad.
Fred.* 10.
Mus cauda longa, rufus, lineis in dorſo al-
bicantibus, margaritarum æmulis. *Briſſon quad.* 124.

R. with round naked ears: of a grey color: the back and ſides
elegantly marked with twelve rows of ſmall pearl-colored
ſpots, extending from the head to the rump: tail the length of
the body: in ſize, half that of a common mouſe.

Inhabits *India*. In the ſame country, and in *Guinea*, is another
very ſmall ſpecies, which ſmells of muſk. The *Portugueſe* living
in *India* call it *Cheroſo*, and ſay its bite is venomous. *Boullaye la
Gouz.* 256. *Barbot's Guinea,* 214.

Mus Barbarus. M. cauda mediocri corpore fufco, ftriis decem pallidis, palmis tri- dactylis, plantis pentadactylis. *Lin. fyft. tom.* i. *pars* ii. *addenda.*

L ESS than the common moufe: of a brown color: marked on the back with ten flender ftreaks: three toes with claws on the fore feet, with the rudiments of a thumb: tail of the length of the body.

Inhabits *Barbary.*

Mus Mexicanus maculatus. *Seb. Muf.* 74. *tab.* xlv. *fig.* 5.

R of a whitifh color, mixed with red: head whitifh: each fide of the belly marked with a great reddifh fpot.

According to *Seba* inhabits *Mexico.*

Mus agreftis Americanus albus. *Seb. Muf.* i. 76. *tab.* xlvii. *fig.* 4.

R with pointed ears and nofe; the laft black: whifkers long: fur very fhort: limbs very weak and flender: tail at the bafe thick, growing gradually fo from the rump, fo that the junction cannot be diftinguifhed; decreafes gradually, and becomes very long and flender; ends in a point, and is in all parts befet with long hair.

Color of this animal univerfally white.

According to *Seba,* found in *Virginia.* The thicknefs at the bafe of the tail is its fpecific difference.

Mus

Mus Vagus. *Pallas Nov. fp. fafc.* i. 327. *tab.* xxii. *fig.* 2.

R. with an oblong head: blunt nofe, with a red tip: cutting
teeth yellow; the upper truncated: eyes placed midway
between the nofe and the ears: ears large, oval, naked; the tip
dufky and downy: limbs flender: inftead of a thumb, on the fore
feet, is a conic wart: tail longer than the body, and very flender.

Color above a pale afh, mixed and undulated with black:
along the back to the tail is a black line: ends of the limbs
whitifh.

Length, from nofe to tail, between two and three inches; the
tail near three.

Inhabits the whole *Tartarian* defert; and at certain times
wanders about in great flocks, and migrating from place to
place during night. Obferved as high as lat. 57, about the *Irtifh*,
and between the *Oby* and *Jenefei*, in birch woods: is of a very
chilly nature; foon becomes torpid, and fleeps rolled up in the
cold night, even of the month of *June*. Lives in fiffures of
rocks, under ftones, and in hollow fallen trees: has carnivorous
inclinations; for on being placed in a box with a moufe of another fpecies, it killed and devoured part, notwithftanding it had
feeds to feed on. Is called by the *Tartars, Dfhickis-fitfkan*, or *gregarious Moufe*.

Mus Betulinus. *Pallas Nov. sp. fasc.* i. 322. *tab.* xxii. *fig.* 1.

R. with a sharp nose, with the end red : ears smaller than those of the former, brown, bristly at the end : limbs very slender : toes long, slender, and very separable : tail very long and slender, much exceeding the length of the body; brown above, white below.

Color of the head and body a cinereous rust, with a few dusky hairs interspersed : breast and belly, pale ash : along the top of the back is a dusky line.

Less than the former.

Inhabits the birch woods about the plains of *Ischim* and *Baraba*, and between the *Oby* and *Jenesei* : lives in the hollows of decayed trees : very tender, and soon grows torpid in cold weather : runs up trees, and fastens to the boughs with its tail; and, by assistance of its slender fingers, adheres to any smooth surface : emits a weak note.

Mus Agrarius. *Pallas nov. sp fasc.* i. 341. *tab.* xxiv. A. *Itin.* i. 454.
Mus Rubeus. *Schwenkfeldt Anim. Silef.* 114.

R. with a sharp nose : oblong head : small ears lined with fur : color of the body and head ferruginous, with a dusky line along the back : belly and limbs whitish : above each hind foot is a dusky circle.

A little less than the field mouse. The tail only half the length of the body.

4

Inhabits

PLACE.   Inhabits the temperate tract of *Ruſſia*, and *Siberia*, as far as the *Irtiſh*: in the former, chiefly about villages and corn-fields; in the latter, in woods. In *Ruſſia* is often migratory, and often very noxious to the grain: it is called there *Shitnik*, or the *Corn Mouſe*, for it abounds in the ſheafs and ricks. At times they wander in vaſt multitudes, and deſtroy the whole expectations of the farmer. This plague did in particular, in the years 1763 or 1764, make great ravages in the rich country about *Caſan* and *Arſk*; but came in ſuch numbers as to fill the very houſes, and became through hunger ſo bold as to ſteal even the bread from the table before the very faces of the gueſts. At approach of winter they all diſappeared.

They make their retreats a little below the ſurface, which in thoſe places appears elevated: each has a long gallery, with a chamber at the end, in which they place their winter food, which conſiſts of various ſorts of ſeeds.

392. SORICINE.   Mus Soricinus. *Schreber*, tab. clxxxiii. *Gm. Lin.* 130.

R. with an elegant ſlender head: ears rounded and covered with hair: tail long and ſlender: hair on the head and upper part of the body cinereous, mixed with yellow: belly white: length two inches.

PLACE.   Inhabits the neighborhood of *Straſburg*: diſcovered by Profeſſor *Herman*.

Mus

Mus pumilio. *Gmel. Lin. ſyſt.* 130. *Sparman's voy.* ii. 349. *tab.* vii.

R. with black forehead and hind part of the head: from the laſt extend along the back to tail four black lines: color of the reſt of the animal a cinereous brown: tail of a light color, very ſmall: not ſuperior in ſize to the following.

Inhabits the foreſt of *Sitſicamma* on the *Slangen* river, at a vaſt diſtance to the eaſtward of the *Cape* of *Good Hope*.

PLACE.

Mus minutus. *Pallas Nov. ſp. faſc.* i. 345. *tab.* xxiv. B. *Itin.* i. 454.

R. with a ſharpiſh noſe: duſky, with a whitenefs at the corner of the mouth: ears ſmall, half hid in the fur: body more ſlender than that of the common mouſe: tail ſhorter and more ſlender.

Color, a deep tawny above, white below: feet grey.

The leſt of the genus; little more than two inches long from noſe to tail; weight not half a dram.

SIZE.

Inhabits the temperate parts of *Ruſſia* and *Siberia*, in corn-fields and barns; is alſo plentiful in the birch-woods. More males among them than females. Seem to wander without any certain places for their neſts.

PLACE.

** With

** With tails of middle length.

Mus Saxatilis. *Pallas Nov. sp. fasc.* i. 255. *tab.* xxiii. B.

R with an oblong head; nose rather pointed: ears rising above the fur; oval, downy, at the edges brown: whiskers short: limbs strong: tail not half so long as the body, with a few hairs scattered over it.

Color above, brown slightly mixed with grey: sides incline more to the last color: belly of a light cinereous: snout dusky, surrounded with a very slender ring of white.

SIZE.   Length four inches: tail one and a half.

PLACE.   Inhabits the country beyond lake *Baikal*, and the *Mongolian* desert: makes its burrows in a wonderful manner, considering the weakness of its feet, between the crannies of the rocks which had been forced open by violence of frost, or the insinuation of roots of plants: it chuses its habitation amidst the rudest rocks, and lives chiefly on the seeds of *Astragali*. The burrows consist, first-ly, of a large winding oblique passage, through which the animal flings out the earth: secondly, of one or more holes pointing downwards, which likewise wind among the rocks; and at their bottom is the nest, formed of soft herbs.

Viverra

Viverra fafciata. *Gmelin Lin. i. 92.* Chat fauvage, &c. *Sonnerat voy.* ii. 143. tab. lxxix.

R. with fhort pointed ears: fharp nofe: two cutting teeth in each jaw, and fourteen grinders in each: five toes to each foot: claws ftrong and crooked: color grey, tinged on the lower part of the head and neck with red: belly white: back and fides marked with four black lines, commencing near the hind part of the head, and ending at the rump: on each thigh is a bifurcated black ftroke, the forks pointing backwards.

Length two feet; tail nine inches. Inhabits *India.* No further account is given by M. *Sonnerat* of this and the following obfcure fpecies. I place them in this genus, as they have no canine teeth, and only two incifores in each jaw.

Le Zenik des Hottentots. *Sonnerat voy.* Viverra Zenik. *Gmelin Lin. i. 84.* ii. 145. tab. xcii.

R. with fhort ears: very long fharp nofe: two cutting teeth; fixteen grinding teeth: four toes on each foot: claws on the fore feet very long, and almoft ftrait: color of a reddifh grey, ftriped tranfverfely with ten black lines falling from the back almoft to the belly.

Size of a water rat: tail not fo long as the body: of a gilded red on three parts of its length; the reft black.

Inhabits the land of the *Hottentots.*

398. Œconomic.   Mus Œconomus. *Pallas Nov. sp. fasc.* i.   Tegoulichitck. *Descr. Kamtschatka, Engl.*
234. *tab.* xiv. A. *Itin.* iii. 692.   *ed.* 104.

R with small eyes: ears naked, and usually hid in the fur:
limbs strong: teeth very tawny: color black and yellow,
intimately mixed; dusky on the back; from throat to tail hoary;
beneath the hair a dark down; ends of the feet dusky.

Size.   Length four inches and a quarter; of the tail, more than an
inch: in form of body like the meadow mouse, but is rather
longer, and the belly bigger. The females are far superior to the
males in size, as on the former rests the chief labor of providing the
food.

Place.   Inhabits in vast numbers all *Siberia*, especially the eastern parts,
and *Kamtschatka*; and even found within the *Arctic* circle.

Manners.   They are called by Doctor *Pallas*, *Mures Œconomi* or Œconomic Mice, from their curious way of living. They inhabit
damp soils, and shun the sandy; form burrows beneath the upper crust of the turfy ground; and have in them many chambers, and several entrances. Never more than two animals are
found in these extensive nests, and those fondly attached to each
other; sometimes only one inhabits these dwellings, except towards
autumn, when the whole family make it their residence. In the
first they form magazines for winter food, consisting of various sorts
of plants, which they collect in summer with great pains; and in
sunny days draw them out of their nests, in order to give them a
more effectual drying. During summer they never touch their
hoards, but live on berries, and other vegetable productions.

3   Twenty,

1

2

3

1. *Lineated Rat* __ N.º 393.   2. *Œconomic Rat* __ N.º 398.

3. *Talpine* _____ N.º 422.

Twenty, and even thirty pounds of fresh roots, have been found in one hoard. Besides man, these mice find a cruel animal in the wild boars, which ransack the magazines, and devour the little defence-less owners.

They in certain years make great migrations out of *Kamtf-chatka*; they collect in the spring, and go off in incredible multitudes. Like the *Lemmus*, they go on in a direct course, and nothing stops their progress, neither rivers nor arms of sea: in their passage they often fall a prey to the ravenous fishes and birds; but on land are safe, as the *Kamtschatkans* pay a superstitious regard for them; and when they find them lying, weak or half dead with fatigue, on the banks, after passing a river, will give them all possible assistance. They set out on their migration westward. From the river *Pengin* they go southward, and about the middle of *July* reach *Ochotska* and *Judoma*, a tract of amazing extent. They return again in *October*. The *Kamtschatkans* are greatly alarmed at their migrations, as they presage rainy seasons, and an unsuccessful chace; but on their return, expresses are sent to all parts with the good news.

Many fables are related of them, such as that they cover their provisions with poisonous herbs before their migrations, in order to destroy other rats which may attempt to plunder their magazines; and if by chance they should be pillaged, they will strangle themselves through vexation, by squeezing their necks between the forks of shrubs; for this reason the natives never take away all their store, but leave part for their subsistence, or leave in its place some caviare, or any thing that will serve for their support. It is certain that the roots of certain poisonous plants are

C c 2

often

often found in their nefts half eaten: but this is no wonder, as it is well known that divers animals will feed on noxious vegetables, which would prove the certain bane of others.

399. WOOLLY. La Chinchilla. *Molina Chili.* 283. Mus laniger. *Gm. Lin.* 134.

R. with very fmall ears: fhort nofe: tail of a middling length, whole body covered with long wool of exquifite finenefs, grey, and long enough to be fpun. The length of this fpecies is fix inches.

Thefe animals live in fociety under ground, and feed on the bulbous roots of the country. It breeds twice a year, and brings five or fix at a time: it is a very gentle tame animal: very fond of being careffed, and will lie down without fear, by mankind: it is often domefticated. The antient *Peruvians* manufactured many fmall articles from the wool, which they fold at a great price.

400. RED. Mus Rutilus. *Pallas Nov. fp. fafc.* i. 146. *tab.* xiv. B.

R. with the nofe and face very briftly: ears, like thofe of the former, naked, except the tip, on which is a rufty down: tail full of hair: color, from the middle of the forehead, along the back, to the rump, an uniform pleafant tawny red: the fides light grey and yellow: under fide of the body whitifh: feet white: tail dufky above, light below.

SIZE. Length not four inches; tail above one.

Inhabits

Inhabits *Siberia*, from the *Oby* eaftward to *Kamtfchatka*, in woods and mountains; and alfo within the *Arctic* circle. Creeps fometimes into houfes and granaries; lives abroad under logs of wood, or trunks of trees: they wander out the whole winter, and are very lively even amidft the fnows: eat any thing which comes in their way; even flefh.

A variety is found about *Cafan*, a little leffer than the *Siberian* kind, and the tail longer and more flender: the red on the back is not fo much diffufed as in the other. The fame kind has alfo been difcovered in the botanical garden at *Gottengen*.

Mus Alliarius. *Pallas nov. fp. fafc.* i. 252. *tab.* xiv. C.

R. with great open naked ears, very apparently out of the fur: tail clothed with hair: color on the back cinereous, mixed with longer hairs tipped with dufky grey: fides of a whitifh afh: breaft, belly, and feet white: tail marked along the top with a dufky line, the reft white.

Length a little above four inches; tail one and a half.

Inhabits the country about the *Jenefei* and *Lena*: is frequent in the fubterraneous magazines of bulbous roots, efpecially the *Allium angulatum*, or angular garlic, formed by the *Siberian* peafants.

R. with the nofe a little extended; four toes on the fore feet, with a tubercle inftead of a thumb: five toes on the hind feet: round ears covered with fur: tail of a middling length, and hairy:

hairy: color of the upper part of the body yellowish grey: belly white.

PLACE.   Inhabits the neighborhood of *Strasbourg.* Discovered by Professor *Herman.*

*** With short tails.

403. LEMMUS.

Lemmar vel Lemmus. *Olaus magnus de gent. Septentr.* 358.
Leem vel Lemmer. *Gesner quad.* 731.
Mus Norvegicus vulgò Leming. *Worm. Mus.* 321, 325. *Scheffer Lapland,* 136. *Pontop. Norway,* ii. 30. *Strom. Sondmor.* 154. *Raii syn. quad.* 227.
Sable-mice. *Ph. Tr. abridg.* ii. 875.
Cuniculus caudatus, auritus, ex flavo, rufo et nigro variegatus. *Brisson quad.* 100.
Mus Lemmus. M. cauda abbreviata, pedibus pentadactylis, corpore fulvo nigro vario. *Lin. syst.* 80. *Pallas nov. sp. fasc.* i. 186. *tab.* xii. A. & B.
Fial-Mus, Sabell-Mus. *Lappis.*
Lummick. *Faun. Suec.* N° 29.
Le Leming. *De Buffon,* xiii. 314.

R. with two very long cutting teeth in each jaw: head pointed: long whiskers; six of the hairs on each side longer and stronger than the rest: eyes small and black: mouth small: upper lip divided: ears small, blunt, and reclining backwards: fore legs very short: four slender toes on the fore feet, covered with hairs; and in the place of the thumb a sharp claw, like a cock's spur: five toes behind: the skin very thin: the color of the head and body black and tawny, disposed in irregular blotches: belly white, tinged with yellow.

SIZE.   Length, from nose to tail, about five inches: in large specimens a little more: the tail about half an inch. Those of *Russian Lapland* are much less than those of the *Norwegian* or *Swedish.*

PLACE.   Inhabits *Norway* and *Lapland,* the country about the river *Oby,*

1. *Lemmus* _____ Nº 403.

2. A Variety of the same.

Oby, and the north extremity of the *Uralian* chain. They appear in numberlefs troops, at very uncertain periods, in *Norway* and *Lapland*: are the peft and wonder of the country: they march like the army of locufts, fo emphatically defcribed by the prophet *Joel*: deftroy every root of grafs before them, and fpread univerfal defolation: they infect the very ground, and cattle are faid to perifh which tafte of the grafs which they have touched: they march by myriads, in regular lines: nothing ftops their progrefs, neither fire, torrents, lake, or morafs. They bend their courfe ftrait forward, with moft amazing obftinacy; they fwim over the lakes; the greateft rock gives them but a flight check, they go round it, and then refume their march directly on, without the left deviation: if they meet a peafant, they perfift in their courfe, and jump as high as his knees in defence of their progrefs: are fo fierce as to lay hold of a ftick, and fuffer themfelves to be fwung about before they quit their hold: if ftruck, they turn about and bite, and will make a noife like a dog.

They feed on grafs, on the rein-deer liverwort, and the catkins of the dwarf birch. The firft they get under the fnow, beneath which they wander during winter; and make their lodgements, and have a fpiracle to the furface for the fake of air. In thefe retreats they are eagerly purfued by the *Arctic* foxes.

They make very fhallow burrows under the turf; but do not form any magazines for winter provifion: by this improvidence it feems that they are compelled to make thefe numerous migrations, in certain years, urged by hunger to quit their ufual refidences.

They breed often in the year, and bring five or fix young at a time:

time : fometimes they bring forth on their migration; fome they carry in their mouths, and others on their backs.

They are not poifonous, as is vulgarly reported; for they are often eaten by the *Laplanders*, who compare their flefh to that of fquirrels.

Are the prey of foxes, lynxes, and ermines, who follow them in great numbers : at length they perifh, either through want of food, or by deftroying one another, or in fome great water, or in the fea. They are the dread of the country : in former times fpiritual weapons were exerted againft them; the prieft exorcifed, and had a long form of prayer to avert the evil * : happily it does not occur frequently; once or twice in twenty years: it feems like a vaft colony of emigrants, from a nation over-ftocked; a difcharge of animals from the great *Northern* hive, that once poured out its myriads of human creatures upon Southern *Europe*. Where the head-quarters of thefe quadrupeds are, is not very certainly known; *Linnæus* fays, the *Norwegian* and *Lapland Alps*; *Pontoppidan* feems to think, that *Kolens* rock, which divides *Nordland* from *Sweden*, is their native place : but wherever they come from, none return : their courfe is predeftinated, and they purfue their fate.

* *Worm. Muf.* 333. where the whole form is preferved. It was once ferioufly believed that thefe animals were generated in the clouds, and fell in fhowers upon the ground: *Per tempeftates et repentinos imbres e cælo decidant, incompertum unde, an ex remotioribus infulis, et huc vento delatæ, an ex nubibus fæculentis natæ deferantur.* Olaus Magnus de Gent. Septentr. 358.

Mus

Mus torquatus. *Pallas Nov. sp. fasc.* i. 205.

R. with a blunt nose: ears hid in the fur: legs strong and short: soles covered with hair: claws very strong, hooked at the end: the hair on the whole body very fine.

Color of the upper part of the body ferruginous, mixed with grey and yellow; sometimes pale grey, clouded with undulated lines of dusky rust-color: from the ears, down each side of the cheeks, is a bed of the same color, and behind that a stripe of white, so that the neck appears encircled with a collar; behind these again is another bed of the former color.

Length to the tail little more than three inches; of the tail one; at its end is a hard tuft of bristles.

Inhabits the northern parts about the river *Oby*. Makes its burrows, with many passages, beneath the turfy soil. The nests are filled with rein-deer and snowy liverworts. They are said to migrate at the same seasons with the *Lemmus*.

Mus Hudsonius. *Pallas Nov. sp. fasc.* i. 208.

R. with slender brown whiskers: very fine long soft hair: cinereous, tinged with tawny, on the back, with a dusky stripe running along its middle: along each side a pale tawny line: belly pale cinereous: limbs very short: fore feet very strong: the two middle claws of the male very strong, thick, and compressed; divided at the end: those of the supposed females (of the

VOL. II.   D d   lesser

lesser skins) small : tail very short, terminated by some stiff bristles.

Length about five inches. Described by Doctor *Pallas*, from some skins sent to him from *Labrador*, one of which he favored me with.

Mus Lagurus. *Pallas Nov. sp. fasc.* i. 210. *tab.* xiii. A. *Itin.* ii. *App.* 704.

R with a long head, and blunt nose : rough lips, and swelling out : ears short, round, flat, just appearing out of the fur : limbs short and slender : tail the shortest of all the genus, scarcely appearing out of the hairs : fur very soft and full, cinereous on the upper part, mixed with dusky : along the back is a dark line : belly and feet of a pale ash-color.

Length between three and four inches.

Inhabits the country above the *Yaik*, *Irtish*, and *Jenesei*. They love dry soils, but firm ; in which they make burrows with two entrances ; one oblique, leading to the nest, the other perpendicular, but both end at it, or unite beyond ; the nest is formed of grass. Usually the male has a different habitation, but sometimes they live together. When more males than one get together, they fight, and the conqueror devours the vanquished ; the mate of the deceased instantly submits to the embraces of the former, even though pregnant. They are very salacious, and bring their young frequently in the air : they bring six at a time : emit often a musky smell when in heat : the males fight sitting up, and bite very hard, and make a noise by striking their teeth together. They sleep very much, and like the Marmots, rolled up;

up; and, like them, are flow in their motions: are very fond of the dwarf iris, but feed on all forts of feeds: they have alfo carnivorous appetites, for they will devour one another, and even others of different fpecies, of the fame fize with themfelves; for which reafon few other kinds live near them. They migrate in great troops; therefore are called by the *Tartars*, *Dfhilkis-Zizchan*, the *Rambling Moufe*.

Mus focialis. *Pallas Nov. fp. fafc.* i. 218. *tab.* xiii. B. *Itin.* ii. *App.* 705.

R. with a thick head and blunt nofe: whifkers white: ears oval, naked: limbs fhort and ftrong: tail flender: nofe dufky: upper part of the body a light grey; paleft on the fides: fides, fhoulders, and belly, white.

Length above three inches; tail an inch.

Inhabits the *Cafpian* defert, between the *Volga* and the *Yaik*, and the country of *Hircania*. They live in fandy, low, and herby places, in large focieties; and in many places the whole ground is covered with the little hills formed by the earth they caft out of their burrows: the burrows are about a fpan in depth, with eight or more paffages. They are always found to live in pairs, or with a family. They live much on tulip-roots. They rarely appear in autumn, but fwarm in the fpring. They are faid either to migrate or change their places in autumn, or to conceal themfelves among the bufhes; and in the winter to fhelter in hay-ricks. They breed later than other kinds. Are the prey of weefels, fitchets, crows, and vipers.

Mus

R A T.

Mus Gregalis. *Pallas nov. sp.* 238. *Gmel. Lin. syst.* 133.

R. with large thin ears appearing above the fur: whiskers black; hair rough and hard; color above a pale grey: the back darkened with dusky hairs, which gradually decline into the lighter color: body below of a dirty white: the legs stronger, the tail thicker, than in the SOCIAL species: about the size of that kind.

Inhabits *Siberia*, but not like the country beyond the *Oby*: most plentiful about the *Baikal* lake and *Trans-Baikal* region; especially those places which abound most with the *Lilium pomponium* and *allium tenuiffimum*; and *Siberia* and *Hircania*. They collect the roots of these and of the *Trifolium Lupinaftrum*, for winter food. They form their lodge beneath the turf, and have many minute entrances: the earth that they fling out is carefully heaped above their lodge, in form of a hillock, to divert the rain. In this retreat the male, female, and the progeny of one year, reside. This species is never observed to migrate.

Mus agreſtis capite grandi brachiurus. *Raii ſyn. quad.* 218.

Mus terreſtris. M. cauda mediocri ſub-pilofa, palmis ſubtetradactylis, plantis pentadactylis, auriculis vellere brevio-ribus. *Lin. ſyſt.* 82.

Molle. *Faun. ſuec.* N° 31 *.

Mus cauda brevi, pilis e nigricante & ſordidè luteo mixtis in dorſo, & ſatu-

ratè cinereis in ventre veſtitis. *Briſſon quad.* 125.

Le Campagnol. *De Buffon*, vii. 369. *tab.* xlvii.

The ſhort-tailed Field-mouſe. *Br. Zool.* i. N° 31.

Erdzeiſl. *Kramer Auſtr.* 316.

Mus arvalis. *Pallas Nov. ſp. faſc.* i. 78. Lev. Mus.

R. with a large head: blunt noſe: ears ſhort, and hid in the fur: eyes prominent: tail ſhort: color of the head and upper part of the body ferruginous, mixed with black: belly deep aſh-color: feet duſky.

Length, from noſe to tail, ſix inches; tail only one and a half, thinly covered with hair, terminated by a ſmall tuft.

Inhabits *Europe, Siberia* and *Hircania*; alſo in great abundance in *Newfoundland;* where it does much miſchief in the gardens: in *England*, ſeldom infeſts gardens: makes its neſt in moiſt meadows: brings eight young at a time: has a ſtrong affection for them: re-ſides under ground: lives on nuts, acorns, and corn.

* The ſpecies, N° 30. *Faun. ſuec.* deſcribed by the ſtyle of *Mus cauda abbreviata, corpore nigro fuſo, abdomine cinereſcente,* ſeems the ſame with this.

Mus

410. GREGA-
RIOUS.

Mus gregarius. M. cauda corpore triplo
  breviore fubpilofa, corpore grifeo fubtus pedibufque albis. *Lin. fyft.* 84.

R. with a fmall mouth and blunt nofe : ears naked, and appearing above the fur : hair on the upper part of the body
black at the roots and tips, ferruginous in the middle : throat,
belly, and feet whitifh : tail thrice as fhort as the body, covered
with thin white hairs ; the end black and afh-color : is a little
larger than the common moufe.

Inhabits *Germany* and *Sweden :* eats fitting up : burrows, and
lives under ground.

***** Short-tailed.

With pouches in each jaw.

411. HAMSTER.

Hamefter, Cricetus. *Agricola An. Subter.*
  486. *Gefner quad.* 738. *Raii fyn. quad.*
  221. *Meyer An.* i. *tab.* lxxxi. lxxxii.
Skrzeczek, Chomik. *Rzaczinfki Polon.*
  232.
Porcellus frumentarius. *Schwenkfelde Theriotroph.* 118.
Krietfch, Hamfter. *Kramer Auftr.* 317.
  Pallas Nov. fp. fafc. i. 83. *Zimmerman.*
  343. 511.
Mus cricetus. M. cauda mediocri, auriculis rotundatis, corpore fubtus nigro,
  lateribus rufefcentibus maculis tribus
  albis. *Lin. fyft.* 82.
Glis ex cinereo rufus in dorfo, in ventre
  niger, maculis tribus ad latera albis.
  Briffon quad. 117.
Le Hamfter. *De Buffon,* xiii. 117. *tab.*
  xiv. xvi. *Suppl.* iii. 183.
German Marmot. *Syn. quad.* N° 200.
  LEV. MUS.

R. with large rounded ears : full black eyes : color on the
head and back, reddifh brown : cheeks red : beneath each
ear a white fpot, and another behind ; a fourth near the hind
legs :

5

2

1

1. *Hamster* _____ N.° 411.

2. *Black Variety of the same.*

legs: breaft, upper part of the fore legs, and the belly, black: tail fhort, almoft naked: four toes, and a fifth claw, on the fore feet, five behind: about nine inches long; tail three.

The males are always bigger than the females; fome weigh from twelve to fixteen ounces: the females feldom exceed four or fix. They vary fometimes in color. About *Cafan* is found frequently a family entirely black.

Inhabits *Auftria*, *Silefia*, and many parts of *Germany*, *Poland*, and *Ukraine*; in all the fouthern and temperate parts of *Ruffia* and *Siberia*; and even about the river *Jenefei*, but not farther to the eaft. They are alfo found in the *Tartarian* deferts, in fandy foil, difliking moift places. They are very fond of fuch fpots which abound with liquorice, whofe feeds they feed on. They fwarm fo in *Gotha*, that in one year 11,564, in another 54,429, and in a third 80,139 of their fkins were delivered at the *Hotel de Ville* of the capital *, thefe animals being profcribed on account of their vaft devaftations among the corn.

They are very deftructive to grain; eating great quantities, and carrying ftill more to its hoard: within its cheeks are two pouches, receptacles for its booty, which it fills till the cheeks feem ready to burft: the *Germans* therefore fay of a very greedy fellow, *Er frifft vuie ein Hamfter.*

They live under ground; firft form an entrance, burrowing down obliquely: at the end of that paffage the male finks one perpendicular hole; the female feveral: at the end of thefe are formed various vaults, either as lodges for themfelves and young, or ftore-houfes for their food: each young has its different apart-

* *De Buffon, Suppl.* iii. 185. quoted from Mr. *Sulzer.*

ment;

ment; each fort of grain its different vault; the firft they line with ftraw or grafs: thefe vaults are of different depths, according to the age of the animal; a young *Hamfter* makes them fcarcely a foot deep; an old one finks them to the depth of four or five; and the whole diameter of the habitation, with all its communi-cations, is fometimes eight or ten feet.

The male and female have always feparate burrows; for ex-cepting their fhort feafon of courtfhip, they have no intercourfe. The whole race is fo malevolent as to conftantly reject all fociety with one another. They will fight, kill, and devour their own fpecies, as well as other leffer animals; fo may be faid to be carnivorous as well as granivorous. If it happens that two males meet in fearch of a female, a battle enfues; the female makes a fhort attachment to the conqueror, after which the connexion ceafes. She brings forth two or three times in a year, and brings from fixteen to eighteen at a birth. Their growth is very quick; and at about the age of three weeks, the old one forces them out of the burrows to take care of themfelves: fhe fhews little affec-tion for them; for if any one digs into the hole, fhe attempts to fave herfelf by burrowing deeper into the earth, and totally neg-lects the fafety of her brood: on the contrary, if fhe is attacked in the feafon of courtfhip, fhe defends the male with the utmoft fury.

They lie torpid from the firft colds to the end of the winter; and during that time are feemingly quite infenfible, and have the appearance of being dead; their limbs ftiff, and body cold as ice: not even fpirits of wine, or oil of vitriol, poured in to them, can produce the left mark of fenfibility. It is only in places be-yond the reach of the air in which it grows torpid; for the fevereft

cold

cold on the furface does not affect it, as has been proved by experiment.

In its annual revival, it begins firft to lofe the ftiffnefs of its limbs; then breathes deeply, and by long intervals: on moving its limbs, it opens its mouth, and makes a rattle in the throat; after fome days it opens its eyes, and tries to ftand; but makes its efforts like a perfon much concerned in liquor; at length, when it has attained its ufual attitude, it refts for a long time in tranquillity, to recollect itfelf, and recover from its fatigue.

They begin to lay in their provifions in *Auguft*; and will carry grains of corn, corn in the ear, and peas and beans in the pods, which they clean in their holes, and carry the hufks carefully out: the pouches above mentioned are fo capacious as to hold a quarter of a pint *Englifh*. As foon as they have finifhed their work, they ftop up the mouth of their paffage carefully. As they lie torpid during the whole fevere feafon, thefe hoards are defigned for their fupport on their firft retreat, and in the fpring and beginning of the fummer, before they can fupply themfelves in the fields. In winter, the peafants go what they call a *Hamfter-nefting*; and when they difcover the retreat, dig down till they difcover the hoard, and are commonly well paid; for, befides the fkins of the animals, which are valuable furs, they find commonly two bufhels of good grain in the magazine. Thefe animals are very fierce; will jump at a horfe that happens to tread near them, and hang by its nofe, fo that it is difficult to difengage them: they make a noife like the barking of a dog. In fome feafons are fo numerous as to occafion a dearth of corn. Pole-cats are their greateft enemies; for they purfue them into their holes, and deftroy numbers.

It is remarkable, that the hair fticks fo clofe to the fkin, as not to be plucked off without the utmoft difficulty.

412. VORMELA.

In my former edition I fuppofed the *Vormela* of *Agricola* * to have been a variety of this kind. He fays it is lefs; the whole body marked with yellow and tawny fpots; the tail cinereous, and white tipped with black; but as he adds that it is a palm and a half long, I muft refer it to another fpecies, or perhaps genus; for it is not unlikely but that it is the fame with the *Sarmatian Weefel*, N° 239.

413. YAIK.

Mus accedula. *Pallas Nov. fp. fafc.* i. 257. *tab.* xviii. A.
Mus migratorius. *Pallas Itin.* ii. *App.* 703.

R. with a thick fnout: blunt nofe: very flefhy lips: upper lip deeply divided: upper fore teeth fmall, yellow, convex outwards, truncated; the lower flender, pointed: eyes large: ears great, oblongly oval, high above the fur, naked: tail very fhort, cylindrical: color about the face white: upper part of the body of a cinereous yellow, mixed with brown; below of a hoary whitenefs.

SIZE.

Length near four inches.

PLACE AND MANNERS.

Inhabits the deferts about the *Yaik*: runs about during night, when it quits its burrow. It is faid by the *Coffacks* to migrate in great numbers out of the deferts, and to be followed by multitudes of foxes, prefaging a good hunting-feafon: but Doctor *Pallas* doubts whether this fpecies, or any of the pouched kinds,

* De anim. fubter. 486.

5   go

go far from their homes, as thofe receptacles for provifion are calculated only for fhort excurfions.

Mus Phæus. *Pallas Nov. fp. fafc.* i. 261. *tab.* xv. A.

R. with the forehead much elevated: edges of the eyelids black: ears naked, oval, ftanding far out of the fur: tail very fhort, flightly furred: color above, a hoary afh-color, with long dufky hairs, running from the neck, along the middle of the back, to the tail: the fides whitifh: the circumference of the mouth, under fide of the body, and the extremities of the limbs, of a fnowy whitenefs.

Length about three inches and a half.

Inhabits the deferts of *Afracan*, about *Zarizyn*; and is taken in traps frequently in winter, in places near to ftables and out-houfes. It is alfo common among the *Hyrcanian* mountains, about the *Perfian* villages, where it commits great ravages among the rice. It does not grow torpid during winter, as is proved by the ftomachs of fuch which are taken in that feafon, being found full of food.

SIZE.
PLACE.

Mus arenarius. *Pallas Nov. fp. fafc.* i. 266. *tab.* xvi. A. *Itin.* ii. *App.* 704.

R. with a longifh head and fnout, and fharp nofe: the pouches very large: ears great, oval, brownifh: body fhort: nails white: color of the upper part of the body hoary: fides, belly, limbs, and tail, of a pure white.

E e 2   Length

Length near four inches; tail above one.

Inhabits the fandy plains of the *Baraba,* not far from the river *Irtifh.* The males inhabit a very deep burrow, with a fingle en-trance, at the bottom of which is the neft, made of the *Elymus arena-rius,* and other plants : other burrows, perhaps of the females, had three entrances : in another, difcovered in *May,* were five young in three nefts; two were preferved alive ; were untameable, very fierce, and would fling themfelves on their back, and defend themfelves by biting: they went out only in the night, and hid themfelves during day in their fodder.

Mus fongarus. *Pallas Nov. fp. fafc.* i. 269. *tab.* xvi. B. *Itin.* ii. *App.* 703.

R. with a thick head and blunt nofe : ears oval, very thin, ap-pear above the fur, are very flightly cloathed with hoary down : tail very fhort, blunt, thick, and hairy: color above, a cinereous grey, marked along the back, from head to tail, with a black line : fides of the head and body marked with great white fpots in certain parts, running into one another, in others bounded with brown: belly and legs white.

Length three inches.

Inhabits, with the former, the *Baraba,* ufually in the dry fandy faline places : dwells during fummer in the fhallow new-begun burrows; thofe of the females have a very deep oblique paffage at the end of it : the neft formed of herbs ; in one of which were feven young; from this neft ran another deep hole, perhaps the winter retreat. The young were much grown, yet blind. Doc-tor *Pallas* preferved them long : they grow foon familiar, contrary

4

to

1 Zaritzyn ——— N.º 414.

2. Songar ——— N.º 416.

to the nature of other mice; would feed from his hand, lap milk, and when placed on a table, fhew no defire of running away; but were flower in all their motions than the other fpecies. They wafhed their faces with their paws, and eat fitting up: wandered about in the day and morning: flept all night rolled up: feldom made any cry, and when they did, it was like that of a bat.

Mus furunculus. *Pallas Nov. fp. fafc.* i. 273.	
Mus Barabenfis. *Itin.* ii. *App.* 704.	417. BARABA.

R. with a fharp nofe: large broad naked ears, dufky edged with white: tail longer than that of the preceding: color of the upper part of the body cinereous yellow, growing paler towards the fides: below of a dirty white: from the hind part of the neck extends a black line, reaching not quite to the tail: tail white, marked above with a dufky line.

Length about three inches and a quarter: tail near one inch. SIZE.

Inhabits the fandy plain of *Baraba*, towards the *Ob*; and between the *Onon* and *Argun*, and about the lake *Dalai* in the *Chinefe* empire. Nothing is known of their manners: the fpecimens from whom the defcriptions were formed, were taken running about the fields. PLACE.

The laft divifion of mice is of thofe which lead a fubterraneous life, like the Mole, which I take the liberty of naming

* * *
* * * Mole-

***
*** Mole-Rat.

418. BLIND.

Mus Typhlus. *Pallas Nov. fp. fafc.* i.
Lemni. *Rzaczinfk. Auft. Pclon.* 325. *De Buffon,* xv. 142.
Slepez. *Gmelin Itin.* i. 131. *tab.* xxii.
Spalax microphthalmus. *Gueldenft. Nov.*

Com. *Petrop.* xiv. 411. *tab.* viii. ix.
Mus oculis minutiffimis, auriculis caudaque nullis. *Lepechen. ibid.* 509. *tab.* xv.
Podolian Marmot. *Syn. quad.* N° 204.

R. with a great head broader than the body: not the left aperture for the eyes; yet beneath the fkin are the rudiments of thofe organs, not bigger than the feed of a poppy: no external ears: the end of the nofe covered with a thick fkin: noftrils very remote, and placed below: the mouth gaping, and the teeth expofed: upper fore teeth fhort, lower very long, and none of them hid by the lip; ends quite even: body cylindrical: limbs very fhort: five toes on each foot, all feparated, except by a thin membrane near the bafe: claws fhort: hair univerfally fhort, thick, and very foft; dufky at the bottom, at the ends of a cinereous grey: the fpace about the nofe, and above the mouth, white.

SIZE.

Length between feven and eight inches: weight of a male above eight ounces.

PLACE.

Inhabits only the fouthern parts of *Ruffia,* from *Poland* to the *Volga,* but is not found any where to the eaft of that river; but is very common from the *Syfran* to the *Sarpa:* is frequent along the *Don,* even to its origin, and about the town of *Ræfk,* excepting the fandy parts, for it delights in moift and turfy foils.

It

It lives in great numbers in the fame places with the EARLESS MARMOTS.

It forms burrows beneath the turf for a very confiderable extent, with feveral lateral paffages made in queft of roots, on which it feeds. At the interval of fome yards, there are openings to the furface to difcharge the earth, which forms in thofe places hillocks of two yards in circumference, and of a great height. It works its way with its great teeth, and cafts the earth under its belly with the fore feet, and again behind it, with its hind feet: it works with great agility; and on any apprehenfion of an enemy, it forms inftantly a perpendicular burrow. The bite of this animal is very fevere. It cannot fee its affailant, but lifts up its head in a menacing attitude. When irritated, it fnorts, and gnafhes its teeth, but emits no cry. It often quits its hole, efpecially in the morning, and during the amorous feafon bafks with the female in the fun. It does not appear that it lies torpid during winter, nor whether it lays in provifion for that feafon. It is particularly fond of the bulbous *Chærophyllum*.

The *Ruffians* call it *Slepez*, or the blind: the *Coffacks*, for the fame reafon, ftyle it *Sfochor Nomon*. In *Ukraine*, the vulgar believe that the touch of a hand, which has fuffocated this animal, has the fame virtue in curing the king's-evil, as was once believed to be inherent in the abdicated family of *Great Britain*.

Mus Afpalax. *Pallas Nov. fp. fafc.* i. 165. *tab.* x. *Itin.* iii. 692.
Mus Myofpalax. *Laxman.*

R. with a thick flat head: fhort fnout: blunt nofe, fit for digging: upper fore teeth naked; lower covered with a
moveable lip: no external ears: eyes very fmall, yet vifible, lodged
deeply in their fockets, which are fo minute as fcarcely to admit
a grain of millet: body fhort, and depreffed: limbs very ftrong,
efpecially the fore legs: fore feet large, and adapted for digging;
naked, and furnifhed with five toes, and very long and ftrong
claws, flightly bent, on the three middle: hind feet naked to the
heel; on each are five toes with fmall claws: tail fhort: hair foft,
and loofe: color at bottom dufky, outwardly of a dirty cinereous grey: in fome is a white line on the hind part of the
head.

SIZE.   Differs in fize. Thofe of the *Altaic* chain are near nine inches
from nofe to tail: thofe about lake *Baikal* not fix: the tail of the
former is near two inches long.

PLACE.   Inhabits, firft, the *Altaic* mountains; and again beyond lake
Baikal, and from thence for fome fpace fouthward; but none are
found to the north. In the former it lives on the bulbs of the
Erythronium; in the latter on thofe of the *Lilium Pomponium.*

It burrows like the former, a little below the furface, and
fpreads over an extent of a hundred fathoms; and the direction
it takes is known by the number of hillocks.

Its voice is weak and plaintive. It digs with both nofe and
fore feet; but lefs than the preceding with the teeth: by comminuting the earth, and flinging it up in hillocks, it prepares the
ground

ground for the reception of various kinds of rare feeds; which grow ufually in greater plenty about fuch places than any others.

The *Tangufi*, about lake *Baikal*, call this fpecies *Monon Zokor*, or blind; yet it is not quite deprived of fight. The *Ruffians* ftyle it *Semiunaja Medwedka*, or *Earth Bear*.

R. with a large head: nofe black; end flatted and corrugated: eyes minute, much hid in the fur: no ears: upper teeth one-third of an inch long, fulcated lengthways; lower, one inch and a quarter, expofed to view: legs fhort: on the fore legs are four toes and a thumb, detached and free: inmoft toe the longeft, the others gradually fhorten: on the thumb is a fhort claw; the other claws are very long, and flightly bent: the foles are naked, and diftinguifhed by two great tubercles: hind feet very long, large, and naked, which the animal refts on even to the heel: they have five toes with fhort claws.

Tail compreffed, and covered above and below with fhort hairs: on the fides befet with very long briftles difpofed horizontally.

Color a cinereous brown, paleft on the lower parts.

Length to the tail thirteen inches: tail two.

Inhabits the fandy country near the *Cape of Good Hope*, where it is called *Sand Moll*. It burrows, and flings up hillocks, like the former; and renders the ground fo hollow, as to be very inconvenient to travellers; for it breaks every fix or feven minutes under the horfes feet, and lets them in up to the fhoulders. This animal feeds on the roots of *Ixiæ*, *Gladioli*, *Antholyzæ*, and *Irides*;

420. AFRICAN.

SIZE.
PLACE.

VOL. II.   F f   grows

grows to the fize of a rabbet, and is by fome efteemed a good difh *. This, from its fuperior fize, I fuppofe to be the *Sand Moll* of Mr. *Maffon*.

421. CAPE.   Mus Capenfis. *Pallas Nov. fp. fafc.* i. 172.   xlvi.
           tab. vii.   La Taupe du Cap. *Journal hift. fig.* 64.
           Long-toothed Marmot. *Brown's Zool. tab.*

R. with a blunt nofe: minute round noftrils: eyes fmall, but larger than thofe of the preceding: no ears: upper fore teeth contiguous, truncated; lower, an inch long, not contiguous, bend upwards, excavated on the upper furface: end of the nofe naked and black, the reft white: chin, and lower fides of the cheeks, of the fame color: fpace round the ears and eyes white: on the hind part of the head is a white fpot; reft of the head, cheeks, back, and fides, of a rufty brown, and cinereous: belly cinereous: five flender toes on each foot, furnifhed with fmall claws: tail very fhort, befet with briftles.

SIZE.   Length, from nofe to tail, about feven inches.

Is very common about the *Cape*, and very deftructive to gardens; flings up hillocks, and eats roots of various kinds.

* *Maffon's trav. Ph. Tranf.* lxvi. 304. *De la Caille*, 299.

Mus

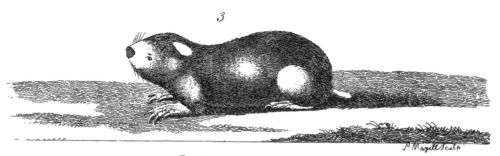

S. Mazell Sculp.

2. *Blind Mole-Rat __ N.°418.*
3. *Daurian _____ N.°419.*
1. *Cape _____ . N.°421.*

Mus Talpinus. *Pallas Nov. fp. fafc.* i. 176. *tab.* xi. B. *Nov. Com. Petrop.* xiv. 568. *tab.* xxi. *fig.* 3.

R. with a large fhort head: thick fnout: nofe truncated: upper teeth extending out of the mouth, long and flat: lower longer, rounded at the ends: eyes fmall, hid in the fur: no ears: the aperture bounded behind by a fmall rim: body fhort: fore feet ftrong: on thofe, and on the hind feet, five toes furnifhed with fmall claws: tail very fhort, fcarcely appearing beyond the fur: color of the head, nofe, back, and fides, dufky: cheeks greyifh: chin white: belly and limbs whitifh.

Length near four inches.

Inhabits all the open grounds and commons of the temperate parts of *Ruffia* and weftern *Siberia*, but fcarcely any beyond the *Irtifh*, and none as far as the *Oby*.

Loves a black turfy foil, and is frequent in meadows near villages: feldom in fandy or mudded tracts: always abound where there is plenty of *Phlomis tuberofa*, and *Lathyrus efculentus*. Its place is known by the little hillocks it flings up along the courfe of its burrow, which is of great extent; for this reafon the *Ruffians* call it *Semleroika*, or *Earth-digger*. In thefe burrows it lurks all the day, but in evening and morning renews its labors; nor does it quit its hole unlefs to fling out the earth, or in the feafon of love to feek a mate, or to change the place of its habitation. It does not bear the full light of day; therefore its few excurfions are ufually in the evenings.

It does not grow torpid in winter; but makes its neft beneath fome fhrub or hay-rick, and deep in the ground, and keep them-

SIZE.
PLACE.

MANNERS.

felves

felves warm by lining it with foft grafs: and often make a lodge, which they fill with tuberous roots. During the cold feafon their fur grows univerfally thicker and longer.

It is very eafily taken: but foon grows fick in confinement, unlefs a quantity of earth is put into the place. They emit a puling note, but that rarely: they often gnafh, and, as it were, whet their teeth againft each other.

They are in heat the end of *March*, or beginning of *April*; at that time the females have a ftrong mufky fmell. They bring three or four at a time.

They fometimes vary in color, and are found quite black.

Two

Two cutting teeth in each jaw, pointing forward.
Long flender nofe: fmall ears.
Five toes on each foot.

Mus aquaticus. *Clufii exot.* 373. *Worm. Muf.* 334.
Mufcovy or Mufk rat. *Raii fyn. quad.* 217. *Nov. Com. Petrop.* iv. 383
Caftor mofchatus. C. cauda longa compreffo-lanceolata, pedibus palmatis. *Lin. fyft.* 79.

Dæfman, *Faun. fuec. No.* 28. *De Buffon,* x. 1.
Caftor cauda verticaliter plana, digitis omnibus membranis inter fe connexis. *Briffon quad.* 92.
Long-nofed Beaver. *Syn. quad.* N° 192.

S. with a long flender nofe, like that of a fhrew-moufe: no external ears: very fmall eyes; tail compreffed fideways: color of the head and back dufky; the belly whitifh afh-color: length, from nofe to tail, feven inches; tail eight.

Inhabits the river *Volga* and lakes adjacent, from *Novogorod* to *Saratof*; never found in *Ruffia*, and its exiftence in *Lapland* doubted[*]. Never goes upon dry land, but wanders from lake to lake, only in fortuitous floods: is often feen fwimming or walking under the water: comes up for air to the furface, or in clear weather fporting on the furface: loves ftagnating waters with high banks, in which it makes burrows twenty feet long: feeds on leeches, and the *larvæ* of water infects: a few fragments of roots have alfo been found in the ftomach. Is not torpid during winter, being often in that feafon taken in nets[†]. Is very flow in its pace: makes holes in the cliffs, with the entrance far beneath the loweft fall of the

[*] Dr. *Pallas*, MSS.   [†] The fame.

water;

water; works upwards, but never to the furface, only high enough to lie beyond the higheft flow of the river: feeds on fifh; is devoured by the Pikes and *Siluri*, and gives thofe fifh fo ftrong a flavor of mufk, as to render them not eatable: has the fame fcent as the former, efpecially about the tail: out of which is expreffed a fort of mufk, very much refembling the genuine kind *. The fkins are put into chefts among cloaths, to drive away moths †, and to preferve the wearers from peftilence and fevers.

At *Orenburg*, the fkins and tails fell for fifteen or twenty *copecs* per hundred. They are fo common near *Nizney Novogorod,* that the peafants bring five hundred apiece to market, where they are fold for one ruble per hundred. The *German* name for thefe animals is *Biefem-ratze*; the *Ruffian, Wychozhol.*

424. PERFUMING.   Mus Pilorides? *Pallas Nov. fp. fafc.* i. 91.
Mus albus Ceylonicus? *Briffon,* 122. LEV. MUS.

S with a long flender nofe: upper jaw extending far beyond the lower: upper fore teeth fhort: lower long, flender, incurvated: whifkers long and white: eyes fmall: ears tranfparent, broad, and round: hair fhort and clofe, on head and body, of a fine pale cærulean: the belly lighter: feet naked and pink-colored.

Length from nofe to tail near eight inches; tail three and a

* *Schober* in *Muller's Samlung Ruff.* vii. 41. 42.
† *Ritchkoff Orenb. Topogr.* i. 286.

half:

Perfuming Shrew — N.º 424.

half: quite naked, round, thick at the bafe, tapering to a point; and of the fame color with the feet.

Inhabits *Java*, and others of the *Eaft Indian* iflands; eats rice; has fo ftrong a fcent of mufk as to perfume every thing it runs over. I have it from the moft undoubted authority, that it will render the wine in a well-corked bottle not drinkable, by merely paffing over it. Cats will not touch them.

Tucan. *Hernandez Nov. Hifp.* 7.　　Le Tucan. *De Buffon*, xv. 159.　　425. MEXICAN.

S. with a fharp nofe: fmall round ears: without fight: two long fore teeth above and below: thick, fat, and flefhy body: fhort legs, fo that the belly almoft touches the ground: long crooked claws: tawny hair: fhort tail: length, from nofe to tail, nine inches.

Inhabits *Mexico*: burrows, and makes fuch a number of cavities, that travellers can fcarcely tread with fafety: if it gets out of its hole, does not know how to return, but begins to dig another: grows very fat, and is eatable: feeds on roots, kidney-beans, and other feeds. M. *de Buffon* thinks it is a Mole; but by the ears, it fhould be claffed here.

Mus araneus figura muris. *Marcgrave*　La mufaraigne de Brafil. *De Buffon*, xv.　426. BRASILIAN. *Brafil.* 229.　　　　　　160.

S. with a fharp nofe and teeth: pendulous fcrotum: of a dufky color, marked along the back with three broad black ftrokes: length, from nofe to tail, five inches; tail two.

5　　　　　　　　　　　　　　　　　　　　　Inhabits

Inhabits *Brafil:* does not fear the cat: neither does that animal hunt after it.

427. MURINE.   S. murinus. S. cauda mediocri, corpore fufco, pedibus caudaque cinereis, *Lin. fyft.* 74.

S. with a long nofe, hollowed beneath: very long hairs about the noftrils: ears rounded, and rather naked: of an afh-color: body of the fize of a common moufe: tail a little fhorter than the body, and not fo hairy.

Inhabits *Java.*

428. FŒTID.

Μυγαλη. *Ælian hift. An. lib.* vi. *c.* 22.
Μυογαλη. *Diofcorid. lib.* ii. *c.* 42.
Mus araneus. *Agricola An. Subter.* 485. *Gefner quad.* 747.
Mus araneus, mus cæcus. *Gefner icon.* 116.
Mus araneus, Shrew, Shrew-moufe, or hardy Shrew. *Raii fyn. quad.* 233.
Mus araneus roftro productiore Spitfmaus. *Klein quad.* 57. *Kramer Auftr.*
317.
Sorex araneus. S. cauda mediocri, corpore fubtus albido. *Lin. fyft.* 74.
Nabbmus. *Faun. fuec.* No. 24.
Mus araneus fupra ex fufco rufus, infra albicans. *Briffon quad.* 126.
La Mufaraigne. *De Buffon,* viii. 57. *tab.* x.
Shrew-moufe. *Br. Zool.* i. 112.

S. with fhort rounded ears: eyes fmall, and almoft hid in the fur: nofe long and flender, upper part the longeft: head and upper part of the body of a brownifh red: belly of a dirty white: length, from nofe to tail, two inches and a half; tail one and a half.

PLACE.   Inhabits *Europe, Siberia,* and even the *Arctic* flats, and *Kamtfchatka;* it is alfo found about the *Cafpian* fea; lives in old walls,

heaps

heaps of ftones, or holes in the earth: is frequently near hay-ricks, dunghills, and neceffary-houfes: lives on corn, infects, and any filth: is often obferved rooting in ordure, like a hog: from its food, or the places it frequents, has a difagreeable fmell: cats will kill, but not eat it: brings four or five young at a time. The antients believed it was injurious to cattle, an error now detected. There feems to be an annual mortality of thefe animals in *Auguft*, numbers being then found dead in the paths.

Mus araneus dorfo nigro, ventreque albo.   64. *tab.* xi.   **429. WATER.**
  Merret Pinax, 167.   Water Shrew-moufe. *Br. Zool. illuftr.*
Sorex fodiens. *Pallas **.   *tab.* cii. LEV. MUS.
La Mufaraigne d'Eau. *De Buffon*, viii.

S. with a long flender nofe: very minute ears; and within each a tuft of white hairs: very fmall eyes, hid in the fur: color of the head and upper part of the body black: throat, breaft, and belly, of a light afh-color: the feet white: beneath the tail a triangular dufky fpot: much larger than the laft: length, from nofe to tail, three inches three quarters; tail two inches.

Inhabits *Europe* and *Siberia*, as far at left as the river *Jenefei*;   PLACE. long fince known in *England*, but loft till *May* 1768, when it was difcovered in the fens near *Revefby Abby, Lincolnfhire*: bur-rows in the banks near the water; and is faid to fwim under wa-

---

* Doctor *Pallas* favored me with feveral prints of this animal in 1765, but never publifhed them: he difcovered it near *Berlin*: it is called there *Græber*, or The Digger.

ter * : is called by the Fen-men the *Blind Mouſe:* chirrups like
a graſshopper, and its note often miſtaken for one.

430. ELEPHANT.   S with a very long, ſlender and little noſe: the whole animal of a
                 S. deep brown color.

PLACE.   Inhabits the neighborhood of the Cape of *Good Hope:* called the
                 Elephant, from its *proboſcis-like* ſnout: engraven from a drawing by
                 Mr. *Paterſon.*   This animal has been very ill repreſented by *Petiver*
                 in his Gazoph. Dec. iii. tab. xxiii. fig. 9. under the title of *Mus*
                 araneus maximus Capenſis.

                          Sorex marinus. *Gm. Lin.* 114.

431. MARINE.

                 S with elongated ſnout, channel'd below: ears rounded, and
                 S. naked: fur of a duſky color; whiſkers grey: tail a little ſhorter
                 than the body: ſize of the common mouſe.
                 Inhabits *Java.*

432. SURINAM.   S with the upper part of the body bay; the lower pale aſh,
                 S. mixed with yellow: tail one half ſhorter than the body.
                 Inhabits *Surinam.*

                          * *L. Baldner,* iii. 137.
                 8

                                                         Sorex

Elephant Shrew. N.º 430.

Sorex pufillus. *Erxlelen.* 122. *Gm. Lin.* 114.

S. with the body hoary above, cinereous beneath: tail *(fubdifticha)* fhort, and whitifh: length of the body three inches feven lines; tail one inch one line.

Inhabits the north of *Perfia:* burrows and lives below ground.

Sorex minutus. S. roftro longiffimo. *Lin. fyft.* 73.

S. with a head near as big as the body: very flender nofe: broad fhort naked ears: whifkers reaching to the eyes: eyes fmall, and capable of being drawn in: hair very fine and fhining; grey above, white beneath: no tail.

Inhabits *Siberia,* about the *Oby* and near the *Kama:* lives in a neft made of *lichens,* in fome moift place beneath the roots of trees: lives on feeds: digs: runs fwiftly: has the voice of a bat.

Sorex exilis. *Gm. Lin.* 115.

S. with a very long flender nofe: in fhape and color like the FOETID, but paler: the tail very flender near the roots, then fuddenly grows remarkably thick and round; and again grows gradually lefs to the end.

LINNÆUS imagines that the laft is the leaft of quadrupeds. Doctor PALLAS, who communicated this fpecies, thinks this has

a better

a better clame to that title, as its weight is only equal to, or very little above half a drachm.

Is very common between, and about the rivers *Jenefei* and *Oby*.

436. WHITE-TOOTHED.

S. of a dufky cinereous color: belly white: cutting teeth white: tail flender and hairy.

437. SQUARE-TAILED.

S. of a dufky cinereous color: belly paler: cutting teeth brownifh: tail inclines to a fquared form.

This fpecies has no bad fmell.

438. CARINATED.

S. of a dufky cinereous whitifh on the belly, with brownifh fore teeth: a white fpot beyond each eye: tail flender and taper, carinated or ridged below.

439. UNICOLOR.

S. of an uniform dufky cinereous color: bafe of the tail narrow, or compreffed.

PLACE.

The above four fpecies inhabit the neighborhood of *Straf-bourg*, and were difcovered by Profeffor *Herman*.

Long

Long nose: upper jaw much longer than the lower.
No ears.

Fore feet very broad, with scarcely any apparent legs before:
hind feet small.

Talpa. *Agricola An. Subter.* 490. *Gesner quad.* 931. *Klein quad.* 60.
Talpa, the Mole, Mold-warp, or Want. *Raii syn. quad.* 236.
Kret. *Rzaczinski Polon.* 236.
Scheer, Scheer-maufs. Maul-wurf. *Kramer Austr.* 314.
Talpa Europæus. T. caudata, pedibus pentadactylis. *Lin. syst.* 73.
Mullvad, Surk. *Faun. suec. No.* 23. *Br. Zool.* i. 108.
Talpa caudata, nigricans pedibus anticis et posticis pentadactylis. *Brisson quad.* 203.
La Taupe. *De Buffon*, viii. 81. *tab.* xii. LEV. MUS.

M. with very minute eyes, hid in the fur: long snout: six cutting teeth in the upper, eight in the lower jaw, and two canine in each: no external ears, only an orifice: fore part of the body thick and muscular; hind part taper: fore feet placed obliquely, broad, and like hands: five toes, each terminated by strong claws: hind feet very small, with five toes to each: tail short: skin very tough, so as scarcely to be cut through: hair short, close set, softer than the finest velvet: usually black, sometimes spotted * with white; sometimes quite white: length five inches three quarters; tail one.

Inhabits *Europe*, and the temperate or southern parts of *Russia* and *Siberia*, as far as the River *Lena*. The *Siberian* is much larger than the *European* Mole.

* Spotted Mole, *Edw.* 268.

It

It lives under ground; burrows with vaſt rapidity with its fore feet; flings the earth back with its hind feet: has the ſenſe of ſmelling exquiſite, which directs it to its food—worms, inſects, and roots: does vaſt damage in gardens, by flinging up the ſoil and looſening the roots of plants: is moſt active before rain, and in winter before a thaw, worms being then in motion: breeds in the ſpring: brings four or five young at a time: makes its neſt of moſs, a little beneath the ſurface of the ground, under the greateſt hillock: raiſes no hillocks in dry weather, being then obliged to penetrate deep after its prey: makes a great ſcream when taken. *Palma Chriſti* and white *hellebore*, made into a paſte, and laid in their holes, deſtroys them. None in *Ireland*.

β. YELLOW M. in form reſembling the *European*, but larger, being ſix inches two-tenths long; the tail one inch: hair ſoft, ſilky, and gloſſy, of a yellowiſh brown color at the ends; dark grey at the roots; brighteſt about the head; darkeſt about the rump: belly of a deep cinereous brown: feet and tail white.

Inhabits *N. America*. Deſcribed from a ſkin in which the jaws were taken out.

Cape Mole __ N.º 441

Talpa Sibiricus verficolor, *Afpalax* dictus. *Seb. Muf.* i. 51. *tab.* xxxii. *fig.* 4, 5. *Klein quad.* 60.

Talpa Afiatica. T. ecaudata, palmis tri-dactylis. *Lin. fyft.* 73.

Talpa ecaudata, ex viridi aurea, pedibus anticis tridactylis, pofticis tetradactylis. *Briffon quad.* 206.

La Taupe dorèe. *De Buffon*, xv. 145.

Variable Mole. *Brown's Zool.* 118. tab. 44.

441. SIBERIAN.

M. with the nofe fhort and blunt: fpace between the tip, and corner of the mouth covered with pale brown hair: from the corner of the mouth, a broad whitifh bar points upwards along the fides of the head: color of the hair on the upper part of the body varied with gloffy green and copper-color: below is of a cinereous brown: in the upper jaw are two fharp cutting teeth; in the lower the fame, with a fharp canine tooth contiguous to them on each fide.

TEETH.

On the fore feet three toes with vaft claws; that on the outmoft toe exceedingly large: on the hind feet five fmall toes and weak claws: no tail: rump round.

Length four inches.

Inhabits the *Cape of Good Hope*, not *Siberia*, as *Seba* fuppofes: Whether this is the *Bles Moll* of the *Dutch*, which lives in the harder grounds about the *Cape* *, I cannot determine.

SIZE.
PLACE.

* *Maffon's Trav. Ph. Tranf.* lxvi. 305.

Sorex

442. RADIATED.　　　Sorex criftatus. S. naribus carunculatis, cauda breviore. *Lin. fyft.* 73.
　　　　　　　　　　　　　　　　　LEV. MUS.

M. with fmall but broad fore legs; five long white claws on each: nofe long; the edges befet with radiated tendrils: hair on the body dufky, very fhort, fine, and compact; on the nofe longer: the hind legs fcaly: five toes on each foot: length, from nofe to tail, three inches three quarters: tail flender, round, and taper; one inch three-tenths long.

PLACE.　　　Inhabits *N. America.* Forms fubterraneous paffages, in different directions, in uncultivated fields; raifes walks about two inches high and a palm broad: the holes often give way and let in the walkers. Feeds on roots: has great ftrength in its legs.

443. LONG-TAIL-
ED.

M. with a radiated nofe: the fore feet pretty broad, hind feet very fcaly, with a few fhort hairs on them: the claws on the fore feet like thofe of the common Mole; on the hind very long and flender: hair on the nofe and body foft, long, and of a rufty brown color: tail covered with fhort hair; the length two inches; that of nofe and body four inches fix-tenths.

PLACE.　　　Inhabits *N. America.* LEV. MUS.

444. BROWN.　　　Sorex aquaticus. S. plantis palmatis, palmis caudaque breviore albis.
　　　　　　　　　　　　　　　　Lin. fyft. 74: LEV. MUS.

M. with a flender nofe: upper jaw much longer than the lower; two cutting teeth in the upper, four in the lower, the two middle of which are very fmall: no canine teeth: fore
feet

1 Radiated Mole ___ N.º 442.

2 Long tailed Mole ___ N.º 443.

feet very broad: nails long: hind feet fmall; five claws on each: hair very foft and gloffy, brown at the ends, deep grey at the bottom: tail and feet white: length, from nofe to tail, five inches and a half: tail very flender, not an inch long.

Inhabits *N. America:* called there the Brown Mole: fent from *New York* by Mr. *A. Blackburne,* with β. Yellow Mole, and No: 442 and 443. The black and fhining purple *Virginian* Mole, defcribed by *Seba* * as the fame with the common kind, was not among thofe that gentleman favoured us with. *Linnæus* places this, and our radiated Mole, in his clafs of SOREX, or SHREW, on account of the difference of the teeth; but as thefe animals pof- fefs the ftronger characters of the MOLE, fuch as form of nofe and body, fhape of feet, and even the manners, we think them better adapted to this genus than to the preceding.

PLACE.

Talpa rubra Americana. *Seb. Muf.* i. 51. *tab.* xxxii. *fig.* 2.

445. RED.

M. of a cinereous red color: three toes on the fore feet, four on the hind: form of the body and tail like the *European* kind.

According to *Seba,* it inhabits *America;* but he does not in- form us whether it is *North* or *South.*

* I. 51. tab. xxxii. fig. 4.

XXXVI.
HEDGE-HOG.

Five toes on each foot.
Body covered with ſtrong ſhort ſpines.

446. COMMON.

Erinaceus. *Agricola An. Subter.* 481.
Echinus terreſtris. *Geſner quad.* 368.
Echinus ſc. Erinaceus terreſtris. Urchin,
　or Hedge-hog. *Raii ſyn. quad.* 231.
Jez. *Rzaczinſki Polon.* 233.
Acanthion vulgaris noſtras. *Klein quad.* 66.
Igel. *Kramer Auſtr.* 314.
Erinaceus Europeus. E. auriculis rotun-
datis naribus criſtatis. *Lin. ſyſt.* 75.
Igelkott. *Faun. ſuec.* N° 22. *Br. Zool.* i.
　106.
Erinaceus auriculis erectis. *Briſſon quad.*
　128. *Seb. Muſ.* i. 78. *tab.* xlix.
L'Heriſſon. *De Buffon,* viii. 28. *tab.* vi.
Hærbe, vel Ganfud. *Forſkal,* iii. *Lev.*
　Mus.

H. with a long noſe: noſtrils bordered on each ſide with a
looſe flap: ears rounded, ſhort, broad, and naked: eyes
ſmall: legs ſhort, naked, and duſky: inner toe the ſhorteſt:
claws weak: upper part of the face, the ſides, and rump, covered
with ſtrong coarſe hair of a yellowiſh and cinereous color; the
back, with ſtrong ſharp ſpines of a whitiſh color, with a bar of
black through their middle: tail an inch long: length, from noſe
to tail, ten inches.

PLACE.

Inhabits *Europe* and *Madagaſcar* *; is common in many parts
of *Ruſſia,* but ſcarcely or ever found in *Siberia:* is in motion
during night; keeps retired in the day: feeds on roots, fruits,
worms, and inſects: erroneouſly charged with ſucking cows and
hurting their udders: reſides in ſmall thickets, in hedges, and at
the bottom of ditches covered with buſhes; lies well wrapped up in

---

* *Flacourt voy. Madagaſcar,* 152, where they are called *Sora.*

mofs,

mofs, grafs, or leaves, and during winter rolls itfelf up and fleeps out that dreary feafon: a mild and helplefs animal; on approach of an enemy, rolls itfelf into the form of a ball, and is then invulnerable.

Erinaceus Auritus. *Pallas* & *Gmelin*, in *Nov. com. Petrop.* xiv. 519. 573. *tab.* xvi. and xxi. *fig.* 4.   447. SIBERIAN.

H. with the upper jaw long and flender: with very large open oval ears, naked, brown round the edges, with foft whitifh hairs within: tail fhorter than that of the common hedge-hog: upper part of the body covered with flender brown fpines, encompaffed at the bafe, and near the ends, with a ring of white: the limbs and belly cloathed with a moft elegant foft white fur.

Generally much inferior in fize to the common kind; but beyond *Baikal* is found much larger than that fpecies.

Is very common in all the fouthern deferts, from the *Don* to the *Oby*.   PLACE.

Grows very fat: fleeps all the winter, lodged in a hole a few inches deep: lives on infects, even the moft cauftic, and will eat (as experiment has been made) above a hundred Cantharides without any injury: rolls itfelf up, and has all the manners of the common kind.

H h 2   Le

448. ASIATIC.   Le petit Tandrek. *Sonnerat, voy.* ii. 146.   Le Tendrac, et Le Tanrec. *De Buffon,*
                 tab. xcviii.   xii. 438.

H. with a long flender nofe: fhort rounded ears: fhort legs: the body marked longitudinally with five broad lines of black, and the fame of white; which are continued over the fhoulders and thighs: the white marks confift of fhort fpines; the black marks are furnifhed with long loofe hairs, which fall quite to the ground: head and face quite black: no tail: length feven inches. *M. de Buffon* has given the figure of a young one.

The other, or the *Tanrec*, is rather larger: covered with fpines only on the top and hind part of the head, the top and fides of the neck, and the fhoulders; the longeft were on the upper part of the neck, and ftood erect: the reft of the body was covered with yellowifh briftles, among which were intermixed fome that were black, and much longer than the others. Each of thefe animals, which are varieties or young of the fame fpecies, had five toes on each foot.

PLACE.   Inhabit the ifles of *India,* and that of *Madagafcar:* are, when of their full growth, of the fize of * rabbets: grunt like hogs: grow very fat: multiply greatly: frequent † fhallow pieces of frefh or falt water: they burrow on land: lie torpid during fix months, during which time their old hair falls off. Their flefh is eaten by the *Indians,* but is very flabby and infipid.

* *Dutch voy. Eaft Indies,* 203. Thofe in the cabinet of the *French* King were much fmaller; probably young.
† *Cauche voy. Madagafcar,* 53. *Flacourt hift. Madagafcar,* 152.

American

1. Common Hedge-Hog. N.º 446.

2. ♀ Asiatic ——————— N.º 448.

American Hedge-hog. *Bancroft Guiana,*   *Lin. ſyſt.* 75. *Briſſon quad.* 131.   **449.** G U I A N A.
144.   Erinaceus Americanus albus. *Seb. Muſ.*
Erinaceus inauris. E. auriculis nullis.   i. 78. *tab. fig.* 3.

H without external ears, having only two orifices for hearing:
has a ſhort thick head: back and ſides covered with ſhort
ſpines of an aſh-color, tinged with yellow: face, belly, legs, and
tail, covered with ſoft whitiſh hair: above the eyes, of a cheſnut color; the hind part and ſides of the head of a deeper color:
length, from noſe to tail, eight inches: tail ſhort: claws long
and crooked.

Inhabits *Guiana.*   PLACE.

D I V.

DIV. II. SECT. IV.

DIGITATED QUADRUPEDS:

Without Cutting Teeth.

DIV. II. SECT. IV. Digitated Quadrupeds.

XXXVII.
S L O T H.

Without cutting teeth in either jaw.
With canine teeth and grinders.
Fore legs much longer than the hind.
Long claws.

450. THREE-
TOED.

Arctopithecus. *Gefner quad.* 869. *Icon. quad.* 95.
Ignavus five per αντιφρασιν Agilis. *Cluf. exot.* 110. 372.
Ai, five Ignavus. *Marcgrave Brafil.* 221.
Sloth. *Raii fyn. quad.* 245. *Edw.* 310.
Ignavus Americanus, rifum fletu mifcens. *Klein quad.* 43.
Tardigradus pedibus anticis & pofticis tridactylis. *Briffon quad.* 21.

Ai, five Tardigradus gracilis Americanus. *Seb. Muf.* xxxiii. *fig.* 2. *Schreber,* ii. 7. *tab.* lxiv.
Ouaikarè, Pareffeux. *Barrere France Æquin.* 154.
Bradypus tridactylus. B. pedibus tridactylis cauda brevi. *Lin. fyft.* 50.
L'Ai. *De Buffon,* xiii. 44. *tab.* v. vi. *Br. Muf.* LEV. MUS.

S. with a blunt black nofe, a little lengthened: very fmall external ears: eyes fmall, black, and heavy; from the corner of each a dufky line: color of the face and throat a dirty white: hair on the limbs and body long and very uneven, of a cinereous brown color, with a black line along the middle of the back: each fide, about the fhoulders, is dafhed with ruft-color; the reft of the back and limbs fpotted irregularly with black. The young, fuch as I fufpect that to be in the *Britifh Mufeum,* have few or no fpots. Tail fhort, a meer ftump: legs thick, long, and aukwardly

3

Three-toed Sloth — N.º 450.

aukwardly placed: face naked: three toes, and three very long claws on each foot.

It grows, as *Nieuhoff* remarks, to the bulk of a middle-fized fox*.

Inhabits moſt parts of the eaſtern ſide of *South America:* the moſt ſluggiſh and moſt ſlow of all animals; ſeems to move with the utmoſt pain; makes a great progreſs if it can go a quarter of a league in a day†: aſcends trees, in which it generally lives, with much difficulty: its food is fruit, or the leaves of trees; if it cannot find fruit on the ground, looks out for a tree well loaded, and with great pains climbs up: to ſave the trouble of deſcending, flings off the fruit, and forming itſelf into a ball, drops from the branches; continues at the foot till it has devoured all; nor ever ſtirs, till compelled by hunger ‡: its motion is attended with a moſt moving and plaintive cry, which at once produces pity and diſguſt, and is its only defence; for every beaſt of prey is ſo affected by the noiſe, as to quit it with horror ‖: its mouth is never without foam: its note, according to *Kircher,* is an aſcending and deſcending *hexachord*§, which it utters only by night: its look is ſo piteous as to move compaſſion; it is alſo accompanied with tears, which diſſuade every body from injuring ſo wretched a being: its abſtinence from food is remarkably powerful; one that had faſtened itſelf by its feet to a pole, and was ſo ſuſpended croſs two beams, remained forty days without meat,

---

* *Nieuhoff's trav. Churchill's collect.* ii. 18.

† *Gumilla Orenoque,* ii. 13.

‡ *Ulloa's voy.* i. 103.

‖ *Ibid.*

§ *Kircher's Muſurgia,* as quoted by Mr. Stillingfleet, in his miſcellaneous tracts, *p.* 100.

drink, or ſleep * : the ſtrength in its feet is ſo great, that there is no poſſibility of freeing any thing from its claws, which it happens to ſeize on. A dog was let looſe at the above-mentioned animal, when it was taken from the pole; after ſome time the *Sloth* layed hold of the dog with its feet, and held him four days, till he periſhed with hunger †.

451. Two-toed.	Tardigradus Ceilonicus fæmina. *Seb.* *Muſ.* i. *tab.* xxxiv. Bradypus didactylus. Br. manibus didactylis cauda nulla. *Lin. ſyſt.* 51. *Schreber,* ii. 10. *tab.* lxv.	Tardigradus pedibus anticis didactylis, poſticis tridactylis. *Briſſon quad.* 22. L'Unau. *De Buffon,* xiii. 34. *tab.* i. *Br. Muſ.*

S. with a round head : ſhort projecting noſe : ears like the human, lying flat to the head : two long ſtrong claws on the fore feet, three on the hind : hair on the body long and rough; on ſome parts curled and woolly : in ſome, of a pale red above, cinereous below; in others, of a yellowiſh white below, cinereous brown above. No tail. Length of that in the *Britiſh Muſeum* eleven inches : I believe a young one.

PLACE. Inhabits *South America* and the iſle of *Ceylon.* The laſt is ſtrenuouſly denied by M. *de Buffon,* who has fixed the reſidence of this genus to *America* only : but, beſides the authority of *Seba,* who expreſsly ſays his ſpecimen was brought from *Ceylon,* a gentleman, long reſident in *India,* and much diſtinguiſhed in the literary world, has informed me he has ſeen this animal brought from the *Paliacat* mountains that lie in ſight of *Madraſs*; which ſatisfies me that it is common to both continents. Farther enquiry is deſired into the identity of this ſpecies.

* *Kircher.* † *Ibid.*

There

Ursiform Sloth ———— *N.° 452.*

There is reafon to think that it is met with alfo in *Guinea*, or
at left fome fpecies of this genus; for *Barbot* and *Bofman* defcribe an animal by the name of *Potto*, to which they give the attributes of the former, and defcribe as being grey when young,
red, and covered with a fort of hair as thick fet as flocks of
wool. Both thefe writers were fenfible men, and, though not naturalifts, were too obfervant of the animals of *Guinea* to miftake
one whofe charaƈters are fo ftrongly marked as thofe of the
Sloth *.

Bradypus urfiformis. *Naturalifts Mifcellany.* tab. 58.   452. URSIFORM.

S. with a long and ftrong nofe, truncated at the end : the forehead rifes fuddenly above it : that and the nofe whitifh, and
almoft naked : eyes very fmall ; above is a black line : ears fhort,
and loft in the hair : the hair on the top of the head points forward, that in the neck is parted in the middle ; on head and neck,
back and fides, is extremely long, fhaggy and black ; in moft parts   HAIR.
twelve inches long, and on the upper part of the body fhines in
the fun with a moft brilliant purple glofs ; on the breaft and
belly fhort ; acrofs the firft is a line of white : the tail is only five
inches long, and is quite hid in the hair : the limbs are very ftrong
and bear-like : on each foot are five toes : on thofe of the fore feet
the claws are three inches long, pointing forward, and flightly incurvated ; pointing forward and admirably adapted for digging or
burrowing : the claws of the hind feet are very fhort : the bottoms

* *Bofman*, 237.   *Barbot*, 212.
I i 2   of

of the feet are black and naked. This animal wants the *incifores*,

or cutting teeth, above and below. In each jaw are two canine teeth, remote from the grinders : the roof of the mouth is marked with tranfverfe fulci : the tongue is fmooth, and not fo long as the mouth.

The noftrils are tranfverfe, and appear like a narrow flit: the lips are very loofe, and capable of being protruded to a great length, and drawn in again; they ferve the ufe of a hand, and by their means it conveys apples or any fort of food, into its mouth : its

principal food was vegetables, and alfo milk : it was very fond of honey, fugar, and other fweets; but did not willingly eat any animal food.

In its manners it was gentle, and very good natured; it fuffered me to put my hand far down its mouth to examine the infide, and to tumble it up and down, to examine the different parts ; nor did it ever offer to bite : it did no more than emit a fhort abrupt roar when I had provoked it highly.

I clafs it, from the teeth, among the *Bradypi*, or *Sloths*, not from its inactivity, or any of its natural properties : it was neither flow nor languid, but was moderately lively : it appeared to have a habit of turning itfelf round and round, every now and then, as if for amufe-ment, in the manner of a dog about to lie down to fleep: it is faid to have a ftrong propenfity to burrowing; and that it was firft dug out of its retreat by thofe who difcovered it.

It inhabits *Bengal*, and lives in certain fand hills not remote from *Patna*. It was about the fize of a black *American* bear, not half grown. When I faw this animal in 1790 it was between four and five years old, fo probably had attained its full growth.

I faw it in company with the ingenious Doctor *Shaw*, of the

Britifh

Britiſh Muſeum. My figure is copied from his *Naturaliſts* Miſcellany: but it was before engraved by Mr. *Catton* in his book of Quadrupeds. Mr. *Bewick* has alſo given a very good figure of it at p. 266 of his beautiful Hiſtory of Quadrupeds with wooden plates.

Without

Without either cutting teeth or canine teeth.

Head, and upper part of the body, guarded by a cruftaceous covering; the middle with pliant bands, formed of various fegments, reaching from the back to the edges of the belly.

453. THREE-
BANDED.

Tatu apara. *Marcgrave Brafil.* 232. *Raii fyn. quad.* 234.
Armadillo feu Tatu genus alterum. *Cluf. Exot.* 109. *Klein quad.* 48.
Tatu feu Armadillo orientalis. *Seb. Muf.* i. *tab.* xxxviii. *fig.* 2, 3.
Dafypus tricinctus. D. cingulis tribus, pedibus pentadactylis. *Lin. fyf.* 53.
Cataphractus fcutis duobus cingulis tribus. *Briffon quad.* 24.
L'Apar, ou le Tatou a trois bandes. *De Buffon,* x. 206. *Schreber,* ii. 28. *tab.* lxxii. A. lxxvi. *fig.* 1. 2.

A, with fhort but broad rounded ears: the cruft on the head, back, and rump, divided into elegant pentangular tuberculated fegments: three bands in the middle: five toes on each foot: fhort tail.

PLACE AND
MANNERS.

The whole genus inhabits *South America:* the manners of all much the fame: burrows under ground; the fmaller fpecies in moift places, the larger in dry, and at a diftance from the fea: keeps in its hole in the day, rambles out at night: when overtaken, rolls itfelf into the form of a ball, which it does by means of the pliant bands on its middle, and thus becomes invulnerable: when furprized, runs to its hole, and thinks itfelf fecure if it can hide its head and fome part of its body. The *Indians* take it by the tail, when the animal fixes its claws in the earth fo ftrongly that there is no moving it till the *Indian* tickles

it

ıt with a ſtick: is hunted with little dogs, who give notice to
their maſter of its haunts by barking, who digs it out; to take
it out incautiouſly is very dangerous, on account of the ſnakes
that commonly lurk in the burrows. Feeds on potatoes, melons,
and roots, and does great damage to plantations: drinks much:
grows very fat, and is reckoned very delicious eating when young;
but when old, has a muſky diſagreeable taſte: is very numerous;
breeds every month, and brings four at a time: is very inoffen-
ſive *.

Tatou. *Belon obſ.* 211. *Portraits,* 106.
Tatu & Tatu paba *Braſil:* Armadillo
Hiſpanis, Luſitanis Encuberto. *Marc-
grave Braſil.* 131.
Cataphractus ſcutis duobus, cingulis ſex.
Briſſon quad. 25.

Daſypus ſex cinctus. D. cingulis ſenis,
pedibus pentadactylis. *Lin. ſyſt.* 54.
L'Encourbert, ou Le Tatou a ſix bandes.
De Buffon, x. 209. *tab.* xlii. *Supplem.*
iii. 285. *tab.* lvii. *Schreber,* ii. 31. *tab.*
lxi. B. Lev. Mus.

454. SIX-BANDED.

A with the cruſt of the head, ſhoulders, and rump, formed of
angular pieces: the bands on the back ſix; between which,
alſo on the neck and belly, are a few ſcattered hairs; tail not the
length of the body, very thick at the baſe, tapering to a point:
five toes on each foot.

Inhabits *Braſil* and *Guiana.*

PLACE.

* The authorities for the natural hiſtory: *Marcgrave,* 231. *Dampier,* ii. 61.
Gumilla Orenoque, iii. 223 to 226. *Nieuhoff,* 19. *Bancroft's Guiana,* 145. *Rochefort
Antilles,* i. 286.

Ayotochtli?

455. EIGHT-
BANDED.

Ayotochtli? *Hernandez Mex.* 314.
Tatuete *Brasiliensibus*, Verdadeiro *Lusi-
tanis. Marcgrave Brasil.* 231. *Clus.
exot.* 330.
Cataphractus scutis duobus cingulis octo.
Brisson quad. 26.
Erinaceus loricatus cingulis septenis pal-
mis tetradactylis, plantis pentadac-
tylis. *Amœn. Acad.* i. 560.
Dasypus septem cinctus. *Lin. syst.* 54.
Le Tatuete, ou Tatou a huit bandes. *De
Buffon*, x. 212. *Schreber*, ii. 34. 36.
tab. lxxii. lxxvi. *fig.* 3, 4.

A. with upright ears, two inches long: small black eyes: eight
bands on the sides: four toes on the fore feet, five on the
hind: length, from nose to tail, about ten inches; tail nine.

PLACE.
Inhabits *Brasil.* Reckoned more delicious eating than the
others.

456. NINE-
BANDED.

Armadillo. *Worm. Mus.* 335.
Tatu porcinus, Schildverkel. *Klein quad.*
48.
Pig-headed Armadillo. *Grew's rarities*,
18. *Raii syn. quad.* 233.
Tatu five Armadillo Americanus. *Seb.
Mus. tab.* xxix. *fig.* 1.
Dasypus novem cinctus. D. cingulis novem, palmis tetradactylis, plantis pentadactylis. *Lin. syst.* 54. *Phil. trans.*
liv. 57. *tab.* vii.
Cataphractus scutis duobus, cingulis novem. *Brisson quad.* 27.
Le Cachichame, ou Tatou a neuf bandes.
De Buffon, x. 215. *tab.* xxxviii. *Supplem.* iii. 287. *tab.* lviii. *Schreber*, i.
37. *tab.* lxxiv. lxxvi. *fig.* 7. 10.
American Armadillo. *Phil. Trans.* liv.
57. *tab.* vii. LEV. MUS.

A. with long ears: crust on the shoulders and rump marked
with hexangular figures; the crust on the head marked in
the same manner: nine bands on the sides, distinguished by trans-
verse cuneiform marks: breast and belly covered with long hairs:
four toes on the fore feet, five on the hind: tail long and taper:
length of the whole animal three feet: the tail a little longer
than the body.

In

Twelve banded Armadillo — N.º 457.

In the Leverian Museum is a specimen of the same form, number of bands, and proportions, with this; but the crusts on the head, and other parts, are covered with large scales not angular.

Inhabits *South America*.   One was brought a few years ago to *England*, from the *Mosquito* shore, and lived here some time: it was fed with raw beef, and milk, but refused our grains and fruit *.

Tatu five Armadillo Africanus. *Seb.*
Muſ. i. *tab.* xxx. *fig.* 3, 4.
Le Kabaſſou, ou Tatou a douze bandes.
De Buffon, x. 218. *tab.* xl.

Cataphractus ſcutis duobus, cingulis duodecim. *Briſſon quad.* 27. *Schreber*, ii. 40. *tab.* lxxv. lxxvi. *fig.* 11. 12.

457. Twelve-banded.

A. with broad upright ears : the crust on the shoulders marked with oblong pieces; that of the rump with hexangular: twelve bands on the sides : five toes, with very large claws, on the fore feet ; five leſſer on the hind: tail shorter than the body : some hairs scattered over the body.

M. *de Buffon* † mentions another of twelve bands, with a tail covered with rhomboid figures, which he is doubtful whether to refer to this species.   It is the largest I ever heard of, being from noſe to tail two feet ten inches long; the tail about one foot eight: by the figure (for I never ſaw the animal) it varies greatly from the other.

* This corroborates what *Marcgrave* ſays of one of these animals, *Cuniculos, aves mortuas aliaque devorant* ; which is very extraordinary in quadrupeds which want both cutting and canine teeth.

† *P.* 256. *tab.* xli.

458. EIGHTEEN-
BANDED.

Weefle-headed Armadillo. *Grew's rarities*, 19.
Tatu Muftelinus. *Raii fyn. quad.* 235.
Dafypus unicinctus. D. tegmine tripartito, pedibus pentadactylis. *Lin. fyft.* 53.

Cataphractus fcuto unico, cingulis octodecim. *Briffon quad.* 23.
Le Cirquinçon, ou Tatou a dixhuit bandes. *De Buffon*, x. 220. *tab.* xlii. *Schreber*, ii. 42.

A with a very flender head: fmall erect ears: the cruft on the fhoulders and rump confifting of fquare pieces: eighteen bands on the fides: five toes on each foot: length, from nofe to tail, about fifteen inches; tail five and a half.

PLACE.   Inhabits *South America.*

DIV.

DIV. II. Sect. V.

DIGITATED QUADRUPEDS:

Without Teeth.

DIV. II. Sect. V. Digitated Quadrupeds.

XXXIX. MANIS.

Back, fides, and upper part of the tail, covered with large ftrong fcales.

Small mouth: long tongue: no teeth.

459. Long-tail-ed.

Lacertus peregrinus fquamofus. *Cluf. exot.* 374. *Raii fyn. quad.* 274.
Scaly Lizard. *Grew's rarities.* 46.
Manis tetradactyla. M. pedibus tetra-dactylis. *Lin. fyft.* 53. *Schreber,* ii. 23. tab. lxx.

Pholidotus pedibus anticis et pofticis te-tradactylis, fquamis mucronatis, cauda longiffima. *Briffon quad.* 19.
Le Phatagin. *De Buffon,* x. 180. *tab.* xxxiv. *Afh. Muf.* Lev. Mus. Br. Mus.

M. with a flender nofe; that and the head fmooth: body, legs, and tail, guarded by large fharp-pointed ftriated fcales: the throat and belly covered with hair: fhort legs: four claws on each foot, one of which is very fmall: tail a little taper, but ends blunt. The color of the whole animal, chocolate.

PLACE.

Inhabits the iflands of *India.* Thefe animals approach fo nearly the genus of Lizards, as to be the links in the chain of beings which connect the proper quadrupeds with the reptile clafs.

They grow to a great length: that which was preferved in the *Mufeum* of the *Royal Society,* was a yard and a half long*: from the tip of the nofe to the tail, was only fourteen inches; the tail itfelf a yard and half a quarter.

* *Grew.*

Long tailed Manis____ N.º 459.

Lacertus fquamofus. *Bontius Java,* 60. *Pet. Gaz. tab.* xx. *fig.* 11.
Armadillus fquamatus major. *Ceilanicus,* feu Diabolus *Tajovaricus* dictus. *Seb. Muf.* i. *tab.* liii. liv. *Klein quad.* 47. *Schreber,* ii. 22. tab. lxix.

Pholidotus pedibus anticis et pofticis pentadactylis, fquamis fubrotundis. *Briffon quad.* 18.
Manis pentadactyla. *Lin. fyf.* 52.
Le Pangolin. *De¹ Buffon,* x. 180. *tab.* xxxiv. *Afb. Muf.* Lev. Mus. Br. Mus.

M. with back, fides, and legs, covered with blunt fcales, with briftles between each: five toes on each foot: tail not longer than the body: ears not unlike the human: chin, belly, and infide of the legs, hairy: tail broad; much fhorter in proportion to the body than that of the preceding, and obtufe at the end: the color of the whole animal a pale yellow.

Inhabits the iflands of *India,* and that of *Formofa.* The *Indians* call it *Pangoelling*; and the *Chinefe, Chin Chion Seick* *. Feeds on lizards and infects: turns up the ground with its nofe: walks with its claws bent under its feet: grows very fat: is efteemed very delicate eating: makes no noife, only a fnorting.

It is alfo found in *Bengal,* where it is called in the *Sanfkrit* language, *Vajracite,* or the *Thunderbolt reptile,* from the exceffive hardnefs of its fcales: in its ftomach is found a number of fmall ftones, probably taken in to help the digeftion. In the fecond volume of the *Afiatic Refearches,* p. 376, publifhed under the direction of the able and learned Sir WILLIAM JONES, is a very good account of this *animal*; under the direction of that gentleman, a

PLACE.

* *Dalhman in Aff. Stock'h.* 1749, 265.

fecond

fecond inundation of knowledge is pouring upon the weftern world from its primeval feat, the *Eaft*.

Perhaps is a native of *Guinea*: the *Quogelo* of the *Negroes*; which *Des Marchais** fays grows to the length of eight feet, of which the tail is four: lives in woods and marfhy places: feeds on ants, which it takes by laying its long tongue crofs their paths, that member being covered with a fticky *faliva*, fo the infects that attempt to pafs over it cannot extricate themfelves: walks very flowly: would be the prey of every ravenous beaft, had it not the power of rolling itfelf up, and oppofing to its adverfary a formidable row of erected fcales. In vain does the leopard attack it with its vaft claws, for at laft it is obliged to leave it in fafety†. The *Negroes* kill thefe animals for the fake of the flefh, which they reckon excellent.

461. BROAD TAILED.

A new *Manis*. Phil. Tranf. vol. lx. p. 36. tab. 11.

M. with five toes on the fore feet, and four on the hind: fcales of the fhape of a mufcle: belly quite fmooth: the exterior fcales end in a fharp point fomewhat incurvated: tail very broad, decreafing to a point: whole length of the animal a *German* ell and five eighths: the tail half an ell and a fpan broad in the broadeft part.

---

* *Voyage du des Marchais*, i. 200. *Barbot*, 114.
† Is faid to deftroy the *Elephant*, by twifting itfelf round the trunk, and compreffing that tender organ with its hard fcales.

This

This ſpecies was found in the wall of a merchant's houſe at *Tran-*
quebar: when purſued it would roll itſelf up ſo that nothing but the
back and tail could be ſeen: it was with great difficulty killed, al-
though it was often ſtruck with rice-ſtampers, or poles armed with
iron: a blow on the belly deprived it of life. The ſcales of this
genus are ſo hard as to ſtrike fire.

Body

XL.
ANT-EATER.

Body covered with hair.

Small mouth: long cylindric tongue.

No teeth.

462. GREAT.

Tamandua-guacu. *Marcgrave Brafil.*
225.
Tamandua-guacu five major. *Pifo Brafil.*
320.
Pifmire-eater, *Nieuhoff*, 19.
Tamandua major cauda panniculata.
Barrere France Æquin. 162.
Mange-fourmis. *Des Marchais,* iii. 307.
Great Ant-Bear. *Raii fyn. quad.* 241.
Myrmecophaga roftro longiffimo, pedibus anticis tetradactylis, pofticis pentadactylis, cauda longiffimis pilis veftita. *Briffon quad.* 15.
Myrmecophaga jubata. M. palmis tetradactylis, plantis pentadactylis. *Lin.*
fyft. 52. *Klein quad.* 45. *tab.* v.
Le Tamanoir. *De Buffon,* x. 141. *tab.*
xxix. *Suppl.* iii. 278. *tab.* lv. *Schreber,*
ii. 14. tab. lxvii. *Br. Muf.*

A. E. with a long flender nofe: fmall black eyes: fhort round
ears: flender tongue, two feet and a half long, which lies
double in the mouth: legs flender: four toes on the fore feet,
five on the hind: the two middle claws on the fore feet very
large, ftrong, and hooked: the hair on the upper part of the
body is half a foot long, black mixed with grey: from the neck,
crofs the fhoulders, to the fides, is a black line bounded above
with white: the fore legs are whitifh, marked above the feet
with a black fpot: the tail is cloathed with very coarfe black
hairs a foot long: length, from nofe to tail, about three feet ten
inches; the tail two and a half: weight about a hundred pounds.

PLACE AND
MANNERS.

Inhabits *Brafil* and *Guiana:* runs flowly: fwims over the great
rivers; at which time it flings its tail over its back: lives on
ants; as foon as it difcovers their nefts, overturns them, or digs
them

them up with its feet; then thrusts its long tongue into their retreats, and penetrating all the paffages of the neft, withdraws it
into its mouth loaded with prey: is fearful of rain, and protects
itfelf againft wet by covering its body with its long tail. This
(as well as every fpecies of this genus) brings but one young at
a time, at which feafon it is dangerous to approach the place:
it does not arrive at its full growth under four years. The flefh
has a ftrong difagreeable tafte, but is eaten by the *Indians*. Notwithftanding this animal wants teeth, it is fierce and dangerous;
nothing that gets within its fore feet can difengage itfelf. The
very Panthers of *America* * are often unequal in the combat; for
if the Ant-eater once has opportunity of embracing them, it fixes
its talons in their fides, and both fall together, and both perifh;
for fuch is the obftinacy and ftupidity of this animal, that it will
not extricate itfelf even from a dead adverfary †: fleeps in the
day; preys by night.

The following hiftory of this animal is given in *Dillon*'s Travels
through *Spain*, p. 76, in his account of the Royal Cabinet of
Natural Hiftory at *Madrid*. " The Great Ant-bear from *Buenos
Ayres*, the *Myrmecophaga Jubata* of *Linnæus*, called by the *Spaniards*
Ofa Palmera, was alive at *Madrid* in 1776, and is now ftuffed and
preferved in this cabinet. The people who brought it from *Buenos
Ayres* fay, it differs from what *they* call the Ant-eater, which only
feeds on emmets, and other infects; whereas this would eat flefh,
when cut in fmall pieces, to the amount of four or five pounds.
From the fnout to the extremity of the tail, this animal is two
yards in length, and his height is about two feet: the head very

---

* *Gumilla Orenoque*, iii. 232.   † *Pifo Brafil.* 320.

narrow; the nose long and slender. The tongue is so singular, that it looks more like a worm, and extends above sixteen inches. His body is covered with long hair, of a dark brown, with white stripes on the shoulders; and when he sleeps, he covers his body with his tail."

The specimen of the Great Ant-eater in the LEVERIAN *Museum*, is superior in size to any we have before heard of.

			Feet.	Inches.
Its whole length is — —	— 7	4
Tail — — —	— 2	9
From tip of the nose to the ears —	— 1	0
Length of the hairs of the mane —	— 1	0
——————————— of the tail — —	— 1	2
Height to the top of the shoulders —	— 2	0

Both of the above are extremely rare, and in an uncommon fine state of preservation.

463. MIDDLE.

Tamandua-i. *Marcgrave Brasil.* 225. *Raii syn. quad.* 242.
Tamandua minor. *Piso Brasil.* 320. *Barrere France Æquin.* 162.
Tamandua-guacu. *Nieuhoff,* 19.
Myrmecophaga rostro longissimo, pedibus anticis tetradactylis, posticis pentadactylis, cauda ferè nuda. *Brisson quad.* 16.
Myrmecophaga tetradactyla. *Lin. syst.* 52. *Zooph. Gronov. No.* 2.
Le Tamandua. *De Buffon,* x. 144. *Schreber,* ii. 16. tab. lxviii.

A. E. with a long slender nose, bending a little down: small black mouth and eyes: small upright ears: bottoms of the fore feet round; four claws on each, like those of the former;

five

five on the hind feet: hair fhining and hard, of a pale yellow color: along the middle of the back, and on the hind legs, dufky: each fide of the neck is a black line, that croffes the fhoulders and meets at the lower end of the back: the tail is covered with longer hair than the back, is taper, and bald at the end: length, from nofe to tail, one foot feven inches; the tail ten inches.

Inhabits the fame country with the laft: its manners much the fame: when it drinks, part fpurts out of the noftrils: climbs trees, and lays hold of the branches with its tail.

PLACE.

Le Tamandua. *De Buffon,* Supplem. iii. 281. tab. lvi.

464. STRIPED.

A. E. with a taper nofe, the upper mandible extending very far beyond the lower: eyes exceedingly fmall: ears round and fhort: tail covered equally with long hairs: five toes on the fore feet.

Body and tail tawny; the firft marked downwards with broad ftripes of black; the laft annulated: legs and nofe ftriped in the fame manner: belly of a dirty white.

Length from nofe to tail thirteen inches *French*; of the tail feven and a half.

M. *de Buffon* fpeaks of one, which he fuppofes to be the fame with this; but the difference in fize and colors forbid us to fubfcribe to his opinion. The account was tranfmitted to him by M. *de la Borde,* phyfician at *Cayenne.* The hair, fays he, is whitifh, and about two inches long: it has very ftrong talons; eats only

in

in the day-time; keeps in the great woods: the flesh is good: it is much more rare than the *great Ant-eater.*

Weighs sixty pounds.

PLACE.

Both these inhabit *Guiana.*

465. LEST.

Tamandua minor flavescens; Ouatiriouaou. *Barrere France Æquin.* 163.	Myrmecophaga didactyla. M. palmis didactylis, plantis tetradactylis, cauda villosa. *Lin. syst.* 51. *Zooph. Gronov.* No. 1.
Tamandua five Coati Americana alba. *Seb. Muf.* i. *tab.* xxxvii	
Myrmecophaga rostro brevi, pedibus anticis didactylis, posticis tetradactylis. *Briffon quad.* 17.	Little Ant-eater. *Edw.* 220.
	Le Fourmillier. *De Buffon,* x. 144. *tab.* xxx. *Schreber,* ii. 17. tab. lxvi.

A. E. with a conic nose, bending a little down: ears small, and hid in the fur: two hooked claws on the fore feet, the exterior much the largest; four on the hind feet: head, body, limbs, and upper part and sides of the tail, covered with long soft silky hair, or rather wool, of a yellowish brown color: from nose to tail seven inches and a half; tail eight and a half, the last four inches of which, on the under side, naked: the tail is thick at the base, and tapers to a point.

Inhabits *Guiana:* climbs trees, in quest of a species of ants which build their nests among the branches: has the same prehensile power with its tail as the former.

PLACE.

There is a fourth species found at the *Cape of Good Hope,* and in *Ceylon;* but being described from a mere fœtus*, we shall avoid giving a transcript of Dr. *Pallas's* account of it, but wait for further information. We shall only say, that it has four toes

* *Pallas Miscel. Zool.* 64.

on

Lest Ant-eater — N.⁰ 465.

on the fore feet, and pendulous ears, which diftinguifhes it from other kinds. *Kolben* * defcribes their manners particularly, and fays they have long heads and tongues, and are toothlefs; and that they fometimes weigh 100lb. †: that if they faften their claws in the ground, the ftrongeft man cannot pull them away: that they thruft out their clammy tongue into the ants neft, and draw it into their mouth covered with infects. That the *African* fpecies agrees with the *American* in every external particular, is confirmed; but that the laft is furnifhed with grinding teeth, like the *Armadillo*, in the lower end of the jaws, is a difcovery proved from the remarks of Doctor *Camper*, a celebrated zoolo-gift in *Holland*. Mr. *Strachan*, in his account of *Ceylon* ‡, gives the fame account of the manners of what the natives call the *Talgoi*, or Ant-Bear. It is not therefore to be doubted, but that thefe animals are common to the old and new continents.

Fourmillier d'Afrique.   *Allamand Suppl.* V. 26. tab. xi.

466. CAPE.

A. E. with a long nofe, truncated at the end like that of a hog; and the noftrils refembling thofe of that animal: ears fix inches long, thin as parchment, and covered with very fine hairs: tongue very long and flender: the hairs on the head and upper part of the body and tail very fhort, and fo clofely adhering to the fkin as if they were glued to it, their color a

DESCRIP.

* *Hift. Cape*, 118; where they are called *Earth Hogs*.
† As quoted by Dr. *Pallas*; I fuppofe from the *Dutch* edition.
‡ *Phil. Tranf. abridg.* v. 180.

dirty

dirty grey; thofe on the fides and belly long and of a reddifh hue; thofe on the legs ftill longer, black and ftraight: the tail thick near the bafe, and tapering to a point: on the fore feet are four toes; on the hind five; all armed with ftrong claws: thofe behind equal even the length of the toes: all are blunted at the end and calculated for burrowing.

SIZE.

The length is three feet five to the origin of the tail, the tail one foot nine.

This fpecies inhabits the neighborhood of the *Cape of Good Hope.*

MANNERS.

It lives under ground; feeds on ants like the other fpecies; but when it has found an ants neft it looks carefully around to fee whether it can feed in fafety, then puts out its long tongue to catch its prey. Is an object of chace among the *Hottentots,* and is reckoned good food.

467. ACULEATED.   Porcupine Ant-eater.   *Naturalift's Mifcellany, pl.* 109.

A E. Length about a foot: coated on the upper parts with fpines refembling thofe of a porcupine, being white tipped with black; the two colors feparated by a ring of tawny or dull orange: fpines on the back and fides fomewhat recumbent, over the tail perpendicularly erect: fnout long, naked, black and tubular, opening very fmall: tongue lumbriciform; forehead, cheeks, and whole under parts of the body, coated with dark brown ftiff hairs: legs very fhort, toes fhort, broad rounded: claws on the fore-feet, five very ftrong, fomewhat obtufe; on the hind-feet four, of which the two firft are much longer, and fharper than the others: thumb unarmed:

PLACE.

tail very fhort. Inhabits *New South Wales:* preys on ants, and is found

Aculeated Ant-eater ____ N.º 467.

found about ant-hills. A moſt extraordinary quadruped, connect-
ing in ſome meaſure the two very diſtant genera of Porcupine,
and Ant-Eater. This ſingular animal is more fully deſcribed by
Dr. *Shaw* in the Naturaliſt's Miſcellany, and from the figure in that
work the repreſentation here given is faithfully copied. Dr. *Shaw*
is of opinion that the genera of Manis, and Myrmecophaga, ought
to be either united, or elſe that this animal ſhould form a diſtinct
genus.

DIV.

DIV. III.

PINNATED QUADRUPEDS;

Having fin-like feet: fore legs buried deep in the ſkin: hind legs pointing quite backwards.

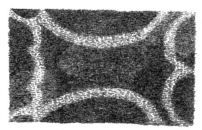

Rubbon Seal — Nº 476.

DIV. III. Pinnated Quadrupeds.

XLI. WALRUS.   With two great tufks in the upper jaw, pointing downwards.
Four grinders on both fides, above and below.
No cutting teeth.
Five palmated toes on each foot.

468. ARCTIC.

Rofmarus. *Gefner Pifc.* 211. *Klein quad.* 92.
Walrus, Mors, Rofmarus. *Worm. Muf.* 289. *Raii fyn. quad.* 191.
Sea-horfe, or Morfe. *Marten's Spitzberg,* 107, 182. *Egede Greenland,* 82.
Sea-cow. *Crantz Greenl.* i. 125. *Schreber,* ii. 88.

Odobenus. La Vache marine. *Briffon quad.* 39.
Trichecus Rofmarus. T. dentibus laniariis fuperioribus exfertis. *Lin. fyft.* 49.
Le Morfe. *De Buffon,* xiii. 358. *tab.* liv. *Br. Muf. Afh. Muf.* Lev. Mus.

W. with a round head: fmall mouth: very thick lips, covered above and below with pellucid briftles as thick as a ftraw: fmall fiery eyes: two fmall orifices inftead of ears: fhort neck: body thick in the middle, tapering towards the tail: fkin thick, wrinkled, with fhort brownifh hairs thinly difperfed: legs fhort; five toes on each, all connected by webs, and fmall nails on each: the hind feet very broad: each leg loofely articulated; the hind legs generally extended on a line with the body: tail very fhort: penis long.

SIZE.   Length, from nofe to tail, fometimes eighteen feet, and ten or

twelve

Arctic Walrus___ N.º 468.

twelve round in the thickeſt part: the teeth have been ſometimes found of the weight * of 20 lb. each.

Inhabit the coaſt of *Spitzbergen, Nova Zembla, Hudſon's Bay,* and
the Gulph of *St. Laurence,* and the *Icy* Sea, as far as Cape *Tſchukiſ-
chi,* and the iſlands off it; but does not extend ſouthward as far as
the mouth of the *Anadyr,* nor are any ſeen in the iſlands between
Kamtſchatka and *America.* Are gregarious: in ſome places appear
in herds of hundreds: are ſhy animals, and avoid places which
are much haunted by mankind †: are very fierce; if wounded in
the water, they attempt to ſink the boat, either by riſing under it;
or by ſtriking their great teeth into the ſides; roar very loud, and
will follow the boat till it gets out of ſight. Numbers of them
are often ſeen ſleeping on an iſland of ice; if awakened, fling themſelves with great impetuoſity into the ſea; at which time it is
dangerous to approach the ice, leaſt they ſhould tumble into the
boat and overſet it: do not go upon the land till the coaſt is
clear of ice. At particular times, they land in amazing numbers:
the moment the firſt gets on ſhore, ſo as to lie dry, it will not
ſtir till another comes and forces it forward by beating it with its
great teeth; this is ſerved in the ſame manner by the next, and
ſo in ſucceſſion till the whole is landed, continuing tumbling
over one another, and forceing the foremoſt, for the ſake of quiet,
to remove further up.

PLACE.

MANNERS.

* Teeth of this ſize are only found on the coaſt of the *Icy* Sea, where the animals
are ſeldom moleſted, and have time to attain their full growth. *Hiſt. Kamtſchatka,*
120.

† In 1608, the crew of an *Engliſh* veſſel killed on *Cherry* Iſle above 900 *Walruſes*
in ſeven hours time; for they lay in heaps, like hogs huddled one upon another.
Marten's Spitzberg. 181, 182.

M m 2　　　　The

The method of killing them on the *Magdalene* iſles, in the gulph of St. *Laurence*, as I am informed, is thus :—The hunters watch their landing, and as ſoon as they find a ſufficient number for what they call a *cut*, go on ſhore, each armed with a ſpear ſharp on one ſide like a knife, with which they cut their throats : great care muſt be taken not to ſtand in the way of thoſe which attempt to get again to ſea, which they do with great agility by tumbling headlong ; for they would cruſh any body to death by their vaſt weight. They are killed for the ſake of their oil, one *Walrus* producing about half a tun. The knowledge of this chace is of great antiquity ; *Octher*, the *Norwegian*, about the year 890, made a report of it to King *Alfred*, having, as he ſays, made the voyage beyond *Norway*, for *the more commoditie of fiſhing of* horſe-whales, *which have in their teeth bones of great price and excellencie, whereof he brought ſome at his returne unto the King* *. In fact, it was in the *northern* world, in early times, the ſubſtitute to ivory, being very white and very hard. Their ſkins, *Octher* ſays, were good to cut into cables. I do not know whether we make any uſe of the ſkin ; but M. *de Buffon* ſays, he has ſeen braces for coaches made of it, which were both ſtrong and elaſtic.

They bring one, or at moſt two, young † at a time : feed on ſea-herbs and fiſh ; alſo on ſhells, which they dig out of the ſand with their teeth : are ſaid alſo to make uſe of their teeth to aſcend rocks or pieces of ice, faſtening them to the cracks, and drawing their bodies up by that means. Beſides mankind, they ſeem to have no other enemy than the white Bear, with whom they have

* *Hakluyt's coll. Voy. i. 5.*   † *Barentz voy. 4.*

terrible

terrible combats; but generally come off victorious, by means of their great teeth.

Le Dugon. *De Buffon*, xiii. 374. *tab*. lvi. *Schreber*, ii. 93.

469. INDIAN.

W. with two ſhort canine teeth, or tuſks, placed in the upper jaw pretty cloſe to each other: in the upper jaw four grinders on each ſide, placed at a diſtance from the tuſks; in the lower, three on each ſide.

Inhabits the *Cape of Good Hope* and the *Philippine* iſles. The head deſcribed above being ſuppoſed to belong to an animal reſembling a *Walrus*, found in the ſeas of *Africa* and *India*, as appears from ſome citations from travellers, too unſatisfactory to merit repetition. It is ſaid by one, that it goes upon land to feed on the green moſs; and that it is called in the *Philippines*, the *Dugung* *.

PLACE.

* *De Buffon*, xiii. 377. *the note.*

Cutting

XLII. SEAL.

Cutting teeth, and two canine teeth in each jaw.
Five palmated toes on each foot.
Body thick at the fhoulders, tapering towards the tail.

470. COMMON.

Φωκη. *Arift. hift. An. lib.* vi. *c.* 12. *Op-pian Halieut.* v. 376.
Vitulus Oceani. *Rondeletii,* 453. 458.
Le Veau Marin, ou Loup de Mer. *Belon Poiffons,* 25.
Phoca. *Gefner Pifc.* 830. *Worm. Muf.* 289. *Klein quad.* 93. *Briffon quad.* 162.
Seal, Seoile, or Sea Calf; Phoca five Vitulus Marinus. *Raii fyn. quad.* 189. *Phil. tranf. abridg. vol.* xlvii. 120. *tab.*

vi. *fig.* 3.
Kaffigiak. *Crantz hift. Greenl.* i. 123.
Phoca vitulina, Ph. capite lævi inauri-culato. *Lin. fift.* 56.
Sial. *Faun. fuec.* N° 4.
Le Phoque. *De Buffon,* xiii. 333. *tab.* xlv. *Schreber,* cxxxiv.
Seal. *Br. Zool.* i. 71, *Br. Zool. illuftr.* xlviii. LEV. MUS.

S. with large black eyes: large whifkers: oblong noftrils: flat head and nofe: tongue forked at the end: two canine teeth in each jaw: fix cutting teeth in the upper jaw; four in the lower: no external ears: body covered with thick fhort hair: fhort tail: toes furnifhed with ftrong fharp claws: ufual length from five to fix feet: color very various, dufky, brinded, or fpotted with white or yellow.

PLACE.

Inhabit moft quarters of the globe, but in greateft multitudes towards the North and the South; fwarm near the *Arctic* circle, and the lower parts of *South America* *, in both oceans; near the

* *Dampier* fays, that they are feen by thoufands on the ifle of *Juan Fernandez*; that the young bleat like lambs; that none are found in the *South Sea, north* of the *equator,* till *lat.* 21; that he never faw any in the *Weft Indies,* except in the *Bay of Campeachy*; nor yet in the *Eaft Indies.* i. 88, 89.

fouthern

fouthern end of *Terra del Fuego*; and even among the floating ice
as low as *fouth lat.* 60. 21 *. Found in the *Cafpian* † Sea, in the
lake *Aral,* and lakes ‡ *Baikal* and *Oron,* which are frefh waters.
They are leffer than thofe which frequent falt waters; but fo fat
that they feem almoft fhapelefs. In lake *Baikal* fome are covered
with filvery hairs; others are yellowifh, and have a large dark-
colored mark on the hind part of the back, covering almoft a
third of the body.

They are found in the *Cafpian* fea, in moft amazing multitudes:
they vary infinitely in their colors: fome are wholly white; others
wholly black; others of a yellowifh white; others moufe colored;
and others again fpotted like a leopard: they creep out of
the fea on the fhores, and are killed as faft as they come; and
are followed by a vaft fucceffion of others, who undergo the
fame fate. It is fingular that the feals of the *Cafpian* are very tenacious of life; it is well known that the fmalleft blow on the nofe
kills thofe of *Europe.* At approach of winter they go up the *Jaik,*
and are killed in great numbers on the ice: they are fought for
the fkins and the oil: numbers are deftroyed by the wolves and
jackals; for which reafon the feal-hunters watch moft carefully the
haunts of the feals in order to drive away their enemies. The feafons
for hunting the feals are fpring and autumn ‖.

Seals bring two young at a time, which for fome fhort fpace
are white and woolly; bring forth in *autumn,* and fuckle their
young in caverns, or in rocks, till they are fix or feven weeks old,

* *Cook's voy.* i. 34.
† *Bell's travels,* i. 49.
‡ The fame, 280.
‖ Decouvertes, &c. faites par les *Ruffes.* ii. 36. 4to. ed.

4

when

when they take to fea: cannot continue long under water; are therefore very frequently obliged to rife to take breath, and often float on the waves. In fummer, fleep on rocks, or on fand-banks: if furprized, precipitate into the fea; or if at any dif-tance, fcramble along, and fling up the fand and gravel with great force with their hind feet, making a piteous moaning: if over-taken, will make a vigorous defence with their feet and teeth: a flight blow on the nofe kills them, otherwife they will bear numbers of wounds. I imagine that the *Cafpian* feal-hunters are not ac-quainted with the method.

Swim with vaft ftrength and fwiftnefs; frolic greatly in their element, and will fport without fear about fhips and boats; which may have given rife to the fable of *Sea-nymphs* and *Sirens*. Their docility is very great, and their nature gentle: there is an inftance of one which was fo far tamed as to anfwer to the call of its keeper, crawl out of its tub at command, ftretch at full length, and return into the water when directed; and extend its neck to kifs its mafter as often and as long as required *.

They never go any great diftance from land: feed on all forts of fifh: are themfelves good food, and often eaten by voyagers: killed for the fake of the oil made from their fat; a young feal will yield eight gallons: their fkins very ufeful in making waift-coats, covers for trunks, and other conveniences: thofe of the lake *Baikal* are fold to the *Chinefe*, who dye, and fell them to the *Mongals* † to face their fur-coats: are the wealth of the *Greenlanders*, fupplying them with every neceffary of life.

* Dr. *Parfons* in *Ph. tranf.* xlvii. 113.
† *Muller's Ruff. Samlung.* iii. 559.

Br.

Pied Seal. — Nᵒ 471.

Br. Zool. i. p. 132.
Le Phoque a ventre blanc. *De Buffon,* Supplem. vi. 310. tab. xliv.

S. with the nose taper and elongated: fore feet furnished with five toes, inclosed in a membrane, but very distinct; the claws long and strait: the hind feet very broad; five distinct toes, with the claws just extending to the margin of the membrane, which expands into the form of a crescent.

This I saw at *Chester*; it was taken near that city in *May* 1766. On the first capture its skin was naked, like that of a porpoise; and only the head, and a small spot beneath each leg, was hairy. Before it died the hair began to grow on other parts: the fore part of the head was black, hind part of the head and the throat white; beneath each fore leg a spot of the same color; hind feet of a dirty white; the rest of the animal of an intense black. I believe they vary in the disposition of the colors: that given by M. *de Buffon* had only the belly white. These species, according to that great writer, frequent the coast of the *Adriatic:* the length of that described by M. *de Buffon* was seven feet and a half; that which I saw was very much less, and probably a young one.

Vitulus Maris Mediterranei. *Rondel.*
Phoca Monachus, capite inauriculato, dentibus incis: utriusque maxillæ qua-tuor, palmis indivisis plantis exunguiculatis. *Herman.*

S. with a small head: neck longer than that of the common seal: orifices of the ears not larger than a pea: hair short

VOL. II. N n and

and rude: color dufky, fpotted with afh-color: above the navel, of the fpecimen defcribed by Mr. *Herman*, was a tawny fpot: the toes on the fore feet furnifhed with nails: the hind feet pinniform, and without nails.

When the animal is placed on its back, the fkin of the neck folds like a monk's hood.

SIZE.

Length of the fpecimen defcribed by Mr. *Herman* was eight feet feven inches: the greateft circumference above five feet.

PLACE.

Inhabits the *Mediterranean Sea*, and as yet not difcovered in the ocean. The common, or oceanic fpecies, is probably an inhabitant of the fame fea, for the fpecies defcribed by *Ariftotle**. is of that kind; he minutely defcribes the feet, and attributes to the hind, as well as the fore feet, five toes, every one furnifhed with nails: that fpecies therefore is the *Phoca* of the antients, not the kind juft under confideration.

473. LONG-NECKED.

Long-necked Seal. *Grew's Mufeum*, 95.

S. with a flender body: length from the nofe to the fore legs as great as from the fore legs to the tail: no claws on the fore feet, which refemble fins.

This was preferved in the *Mufeum* of the *Royal Society*.

Doctor *Parfons* has given a figure of it in the xlviith vol. of *Ph. Tr.* tab. vi. but we are left uninformed of its place.

* *Hift. an. lib.* ii. c. 1.

ALLIED

ALLIED to this is another SEAL in the fame *Mufeum*, fent of late, years from the *Falkland ifles:* its length is four feet: hair fhort, cinereous tipped with dirty white.

Nofe fhort, befet with ftrong black briftles: fhort, narrow, pointed auricles.

Upper cutting teeth fulcated tranfverfely; the lower in an oppofite direction: on each fide of the canine teeth, a leffer, or fecondary one; grinders conoid, with a fmall procefs on one fide near the bafe.

No claws on the fore feet; but beneath the fkin evident marks of the bones of five toes: the fkin extends far beyond their ends. On the toes of the hind legs are four long and ftrait claws; but the fkin ftretches far beyond, which gives them a very pinniform look.

This fpecies probably inhabits alfo the feas about *Juan Fernandez*; for *Don Ulloa** informs us of one kind, which is not above a yard long. The fmall Seals inhabit from the *Falkland Iflands*, round *Cape Horn*, even as far as *New Zealand*; and are feen further from fhore than any other kind. They are very fportive, dipping up and down like porpoifes, and go on in a progreffive courfe like thofe fifh. When they fleep, one fin generally appears above the water. They perhaps extend as far as the *Society Iflands*, at left the natives have a name for the Seal, which they call *Humi*.

* *Ulloa* fays, the firft fpecies of Seal found near that ifle, is not above a yard long. ii. 226.

Tortoife-headed

475. TORTOISE-
HEADED.

Tortoife-headed Seal. *Ph. Tranf.* xlvii. 120. *tab.* vi.

S. with a head like that of a tortoife : neck flenderer than head
or body : feet like thofe of the common Seal.

We are indebted to Doctor *Parfons* for the account of this
fpecies, who fays it is found on the fhores of many parts of
Europe.

476. RUBBON.

S. with very fhort fine gloffy briftly hair, of an uniform color,
almoft black; marked along the fides, and towards the
head and tail, with a ftripe of a pale yellow color, exactly refembling a rubbon laid on it by art; words cannot fufficiently
convey the idea, the form is therefore engraven on the title of
Divifion III. *Pinnated Quadrupeds,* from a drawing communicated
to me by Doctor *Pallas,* who received it from one of the remoteft
Kuril iflands.

Its fize is unknown, for Doctor *Pallas* received only the middle
part, which had been cut out of a very large fkin, fo that no
defcription can be given of head, feet or tail : a. fhews the part
fuppofed to be next-to the head ; b. that to the tail.

OBSCURE SPE-
CIES.

Other obfcure fpecies in thofe feas, which are mentioned in
Steller's MSS. are, I. A middle-fized Seal, elegantly fpeckled in
all parts : II. One with brown fpots, fcarcer than the reft: III. A
black fpecies with a peculiar conformation of the hind legs.

Phoca Leporina. *Lepechin. act. acad. Petrop. pars* i. 264. *tab.* viii. ix.

S. with fur, foft as that of a hare, upright and interwoven; of a dirty white color: whifkers long and thick, fo that the animal appears bearded: head long: upper lip thick: four cutting teeth above; the fame below: nails on fore and hind feet.

Ufual length fix feet and a half; greateft circumference five feet two.

Inhabits the White fea during fummer; afcends and defcends the rivers in queft of prey; found alfo off *Iceland*, and from *Spitfbergen* to the *Tchutkinofs*.

Sea Calf. *Phil. Tranf.* ix. 74. *tab.* v. Le grand Phoque. *De Buffon*, xiii. 345. Utfuk ? *Crantz Greenl.* i. 125. *Schreber Cab.* i. 43. LEV. MUS.

S. refembling the common, but grow to the length of twelve *. feet: that defcribed in the *Phil. Tranf.* was feven feet and a half long, yet fo young. as to have fcarce any teeth; the common Seal is at full growth when it has attained the length of fix.

Inhabits the coaft of *Scotland*, and the fouth of *Greenland*. The fkin is thick, and is ufed by the *Greenlanders* to cut thongs out of for their Seal fifhery. Perhaps is the fame with the great *Kamtfchatkan* Seal, called by the *Ruffians*, *Lachtach*, weighing 800 lb.†, whofe cubs are black.

* A gentleman of my acquaintance fhot one of that fize in the north of *Scotland*.

† *Muller's Voy. Kamtfchatka*, Co.

Neitfek.

479. ROUGH.

Neitſek. *Crantz Greenl.* i. 124. *Schreber*, clxxxvi.

S. with rough briſtly hair, intermixed like that of a hog; of a pale brown color.

PLACE.

Inhabits *Greenland:* the natives make garments of its ſkin, turning the hairy ſide inmoſt. Perhaps what our *Newfoundland* Seal-hunters call *Square Phipper*; whoſe coat, they ſay, is like that of a water-dog, and weighs ſometimes 500 lb.

480. PORCINE.

Phoca porcina. *Molina Chili.* 260.

S. agreeing in general form with the *Urſine*, N° 485, but the noſe is longer, and reſembles a hog's ſnout; it has alſo the veſtiges of ears: the feet have five diſtinct toes, covered with a common membrane.

PLACE.

Inhabits the coaſt of *Chili*, but is a rare ſpecies.

481. EARED.

S. with conoid head: noſe rather pointed: ears an inch long, very narrow and pointed: whiſkers very long and white: fore feet pinniform; neither toes nor nails apparent, terminated membraneouſly: in the hind feet the toes apparent, and each furniſhed with its nail; the membrane extends beyond, and then divides in o five narrow diviſions, correſpondent to each toe: the tail a
little

Harp Seal — N.º 483.

little more than an inch long: the whole body is covered with longish hair of a whitish or creme-color: the length from nose to tail is rather more than two feet.

Inhabits the streights of *Magellan*. This species is finely preserved in Mr. *Parkinson's* Museum, on the southward side of *Black-friars Bridge*. That gentleman has very properly placed seizing on it the *Condor* vulture, the vast co-inhabitant of the *Magellanic* regions. Every one knows that Mr. *Parkinson* is now possessed of the late Sir *Ashton Lever's Museum*; I have therefore still retained the words Lev. Mus. to the description of every animal contained in that matchless collection.

PLACE.

Clap-myss. *Egede Greenl.* 84.   Neitserfoak. *Crantz Greenl.* i. 124.

482. HOODED.

S. with a strong folded skin on the forehead, which it can fling over its eyes and nose, to defend them against stones and sand in stormy weather: its hair white, with a thick coat of thick black wool under, which makes it appear of a fine grey.

Inhabits only the south of *Greenland*, and *Newfoundland:* in the last is called the *Hooded Seal:* the hunters say they cannot kill it till they remove the integument on the head.

PLACE.

Black-sided Seal. *Egede Greenl. plate* iii.   Phoca oceanica. Krylatca Russ. *Lepechin*
Attarsoak. *Crantz Greenl.* i. 124. *Schre-*   *act. acad. Petrop. pars* i. 259. *tab.* vi.
ber, Cab. i. 39.   vii.

483. HARP.

S. with a pointed head and thick body, of a whitish grey color, marked on the sides with two black crescents, the horns
pointing

pointing upwards towards each other; does not attain this mark till the fifth year; till that period, changes its color annually, and is diftinguifhed by the *Greenlanders* by different names each year.

PLACE. Inhabits *Greenland* and *Newfoundland, Iceland,* the *White Sea,* and *Frozen Ocean,* and paffes through the *Afiatic* ftrait, as low as *Kamtfchatka:* is the moft valuable kind; the fkin the thickeft and beft, and its produce of oil the greateft: grows to the length of nine feet. Our Fifhers call this the *Harp,* or *Heart* Seal, and ftyle the marks on the fides the faddle. There is a blackifh variety, which they fay is a young Harp, called *Bedlemer.*

484. LITTLE.   Le petit Phoque. *De Buffon,* iii. 341. *tab.* liii. *Schreber,* cxxxv.
LEV. MUS.

S. with the four middle cutting teeth of the upper jaw bifurcated; the two middle of the lower jaw flightly trifurcated: a rudiment of an ear: the webs of the feet extending far beyond the toes and nails: hair foft, fmooth, and longer than in SIZE.   the common Seal: color dufky on the head and back; beneath brownifh: length two feet four inches.

Our Seal-hunters affirm, that they often obferve, on the coaft of *Newfoundland,* a fmall fpecies, not exceeding two feet, or two feet and a half, in length. M. *de Buffon* fays the fpecimen in the cabinet of the *French* king came from *India*; but from the authority of *Dampier,* and of modern voyagers to the *Eaft Indies,*

who

C

Ursine Seal ___ N.º 485.

who·have affured me they never faw any Seals *·there, I fufpect he was impofed on.

Captain *Abraham Dixon* affured me that he faw off the coaft of *North America,* in his voyages of 1785 to 1788, multitudes of fmall Seals, not exceeding a foot in length : they were perpetually dipping and·rifing again, but were fo active that he never could procure a fpecimen.

Urfus Marinus. *Steller. Nov. Com. Petrop.* ii. 331. *tab.* xv.
Sea Cat. *Hift. Kamtfchatka,* 123. *Muller's Exped.* 59.

Phoca Urfina. Ph. capite auriculato. *Lin. fyft.* 55.
L'Ours Marin. *Briffon quad.* 166. *Schreber,* cxxxii.

485. URSINE.

THERE are three marine animals, which keep a particular fituation, and feem divided between the N. E. of *Afia,* and N. W. of *America,* in the narrow feas between thofe vaft continents. Thefe are what are called the *Sea Lion* and *Sea Bear,* and the *Manati.* They inhabit, from *June* to *September,* the ifles that are fcattered in the feas between *Kamtfchatka* and *America,* in order to copulate, and bring forth their young in full fecurity. They never land upon *Kamtfchatka.* The accurate and indefatigable naturalift *Steller* was the firft who gave an exact defcription of them; he and his companions, in the *Ruffian* expedition of 1742, were in all probability the firft *Europeans* who gave

PLACE.

---

* A gentleman, the moft curious, and greateft navigator of the *Indian* feas now living, informed me, that he not only never met with any Seals in thofe feas, but even none nearer than the ifles of *Gallopagos,* a little north of the line, on the coaft of *America.*

them any difturbance in thofe their retreats. In *September*, thefe animals quit their ftations, vaftly emaciated; fome return to the *Afiatic*, others to the *American* fhores; but, like the Sea Otters, are confined in thofe feas between lat. 50 and 56.

They are not, as far as I can difcover, found from thofe places, any where nearer than *New Zealand* *, where they are very common, and again about *Staten Land* †, the frozen ifland of *New Georgia* ‡, and the *Falkland* iflands ||. I fufpect that they are alfo found in the ifland of *Juan Fernandez*; for, among the Seals fo imperfectly defcribed by Don *Ulloa* §, his fecond kind feems to be of this fpecies. I may add, that *Alexander Selkirk* fpeaks of Seals which come on fhore in that ifland in *November* to whelp ┼, which nearly correfponds with the time our late circumnavigators faw them in *New Year's iflands*, where they found them and their young in *December*. Laftly, I may mention the ifles of *Gallopagos*, where Captain *Woodes Rogers* fays he was attacked by a fierce Seal, as big as a bear, and with difficulty efcaped with his life **.

The *Urfine* Seal, a name we fubftitute for the fea-bear, leads, during the three months in fummer, a moft indolent life: it arrives at the iflands vaftly fat; but during that time they are fcarce ever in motion: confine themfelves for whole weeks to one fpot, fleep a great part of the time, eat nothing, and, except the employment the females have in fuckling their young, are totally inactive. They live in families; each male has from eight to fifty females, whom he guards with the jealoufy of an eaftern monarch; and though they lie by thoufands on the fhores, each family

---

* *Forfter's obf.* 189. † *Cook's voy.* ii. 203. ‡ *Cook's voy.* ii. 213.
Forfter's voy. ii. 529. || *Pernetti, Engl. ed.* 187. *tab.* xvi. § *Voy.* ii. 226.
┼ In *Woodes Rogers's voy.* 136. ** The fame, 265.

4

keeps

keeps itſelf ſeparate from the reſt, and ſometimes, with the young and unmarried ones, amount to a hundred and twenty. The old animals, which are deſtitute of females, or deſerted by them, live apart, and are exceſſively ſplenetic, peeviſh, and quarrelſome: are exceſſively fierce, and ſo attached to their old haunts, that they would die ſooner than quit them. They are monſtrouſly fat, and have a moſt hircine ſmell. If another approaches their ſtation, they are rouzed from their indolence, and inſtantly ſnap at it, and a battle enſues; in the conflict, they perhaps intrude on the ſeat of another: this gives new cauſe of offence, ſo in the end the diſcord becomes univerſal, and is ſpread thro' the whole ſhore.

The other males are alſo very iraſcible: the cauſes of their diſputes are generally theſe:—The firſt and the moſt terrible is, when an attempt is made by another to ſeduce one of their miſtreſſes, or a young female of the family. This inſult produces a combat, and the conqueror is immediately followed by the whole ſeraglio, who are ſure of deſerting the unhappy vanquiſhed. The ſecond reaſon of a quarrel is, when one invades the ſeat of another. The third ariſes from their interfering in the diſputes of others. Theſe battles are very violent; the wounds they receive are very deep, and reſemble the cuts of a ſabre. At the end of a fight they fling themſelves into the ſea, to waſh away the blood.

The males are very fond of their young; but very tyrannical towards the females: if any body attempts to take their cub, the male ſtands on the defenſive, while the female makes off with the young in her mouth; ſhould ſhe drop it, the former inſtantly quits his enemy, falls on her, and beats her againſt the ſtones,

till

till he leaves her for dead. As foon as fhe recovers, fhe comes in the moft fuppliant manner to the male, crawls to his feet, and wafhes them with her tears: he, in the mean time, ftalks about in the moft infulting manner; but in cafe the young one is carried off, he melts into the deepeft affliction, and fhews all figns of deep concern. It is probable that he feels his misfortune the more fenfibly, as the female generally brings but one at a time, never more than two. Even the cubs of thofe on the ifland of *New Georgia* * are very fierce, barking at our failors as they paffed by, and biting at their legs. The breeding-time in this ifland is in the beginning of *January.*

They fwim very fwiftly, at the rate of feven miles an hour. If wounded, will feize on the boat, and carry it along with vaft impetuofity, and oftentimes fink it. They can continue a long time under water. When they want to climb the rocks, they faften with the fore paws, and fo draw themfelves up. They are very tenacious of life, and will live for a fortnight after receiving fuch wounds as would immediately deftroy any other animal.

DESCRIPTION.   The male of this fpecies is vaftly fuperior in fize to the female. The bodies of each are of a conic form, very thick before, and taper to the tail. The length of a large one is eight feet; the greateft circumference five feet; near the tail, twenty inches. The weight 800 lb. The nofe projects like that of a pug dog, but the head rifes fuddenly: noftrils oval, and divided by a *feptum:* the lips thick; their infide red and ferrated: whifkers long and white.

The teeth lock into each other when the mouth is clofed. In

* *Forfter's voy.* ii. 516. 529.

the

the upper jaw are four cutting teeth, each bifurcated; on both fides is a fmall fharp canine tooth bending inwards; near that another, larger: the grinders refemble canine teeth, and are fix in number in each jaw: in the lower jaw are alfo four cutting teeth and two canine: but only four grinders in each jaw: in all, thirty-fix teeth.

Tongue bifid: eyes large and prominent: iris black: pupil fmaragdine: the eyes may be covered at pleafure with a flefhy membrane: the ears are fmall, fharp-pointed; hairy without, fmooth and polifhed within.

The length of the fore-legs is twenty-four inches; like thofe of other quadrupeds, not immerfed in the body like thofe of Seals: the feet are formed with toes, as thofe of other animals, but are covered with a naked fkin, fo that externally they feem a fhapelefs mafs, and have only the rudiments of nails to five la- tent toes: the hind legs are twenty-two inches long, are fixed to the body quite behind, like thofe of Seals, but are capable of be- ing brought forward, fo that the animal makes ufe of them to fcratch its head: thefe feet are divided into five toes, each di- vided by a great web, and are a foot broad: the tail is only two inches long.

The hair is long and rough; beneath which is a foft down, of a bay-color: on the neck of the old males the hair is erect, and a little longer than the reft. The general color of thefe animals is black, but the hairs of the old ones are tipt with grey. The females are cinereous. The fkins of the young, cut out of the bellies of their dams, are very ufeful for cloathing, and coft about 3 s. 4 d. each; the fkin of an old one, 4 s.

The fat and flefh of the old males is very naufeous; but the

flefh.

flefh of the females refembles lamb; and the young ones roafted are as good as fucking-pigs.

486. BOTTLE-NOSE.

Sea Lion. *Dampier's voy.* i. go. iv. 15. *Rogers's voy.* 136. *Anfon's voy.* 122. Phoca Leonina. Ph. capite anticè criftato. *Lin. fyft.* 55.

Le Lion Marin. *Briſſon quad.* 167. *De Buffon*, xiii. 351. *Schreber*, cxxxiii. Le Lame. Phoca elephantina. *Molina Chili.* 261.

S. (*the male*) with a projecting fnout, hanging five or fix inches below the lower jaw: the upper part confifts of a loofe wrinkled fkin, which the animal, when angry, has the power of blowing up, fo as to give the nofe an hooked or arched appearance: the feet fhort and dufky; five toes on each, furnifhed with nails: the hind feet have the appearance of great laciniated fins: large eyes: great whifkers: hair on the body fhort, and of a dun color; that on the neck a little longer: the fkin very thick.

SIZE.

Length of an old male twenty feet; greateft circumference, fifteen.

Female. Nofe blunt, tuberous at the top: noftrils wide: mouth breaking very little into the jaws; two fmall cutting teeth below, two fmall and two larger above; two canine teeth, remote from the preceding; five grinders in each jaw; all the teeth conic: eyes oblique and fmall: auricles none: fore legs twenty inches long: toes furnifhed with flat oblong nails: hind parts, inftead of legs, divided into two great bifurcated fins: no tail:

SIZE.

the whole covered with fhort ruft-colored hair. Length, from nofe to the end of the fins, four yards: greateft circumference two yards and a half*.

* Defcribed from a well-preferved fpecimen in the *Mufeum* of the ROYAL SOCIETY. This is the animal called by Dr. *Parfons*, a *Manati*.

Inhabits

Inhabits the feas about *New Zealand* *, the ifland of *Juan Fernandez* † and the *Falkland iflands* ‡, and that of *New Georgia* ‖, S. *lat.* 54—40. Are feen in great numbers, in *June* and *July*, the breeding-feafon, on the ifland of *Juan Fernandez*, which they refort to for the purpofe of fuckling their young on fhore, and continue there till *September*. They bring two at a time. The female, during that feafon, is very fierce: one of Lord *Anfon's* failors was killed by the enraged dam of a whelp, which he had robbed her of. The male fhews little attachment to its young, but the female is exceffively fond of it: the former will fuffer it to be killed before his face without fhewing any refentment. Towards evening, both male and female fwim a little way to fea, the laft with the young on its back, which the male will pufh off, as if to teach it to fwim.

They arrive on the breeding-iflands very fat and full of blood: when they are in motion, they feem like a great fkin full of oil, from the tremulous movement of the blubber, which has been found to be a foot thick. The *Spaniards* very properly call thefe, and the *Urigne lobos de Aceyte*, or *oil wolves*, from their looking like a fkin full of oil, from the motion of the vaft quantity of fat or blubber, of which their bodies confift §. One has been known to yield a butt of oil; and fo full of blood, that what has run out of a fingle animal ‡ has filled two hogfheads. The flefh is eatable: Lord *Anfon's* people eat it under the denomination of beef, to diftinguifh it from that of Seal, which they called lamb.

---

* *Forfter's obf.* 190. † *Anfon's voy.* 122. ‡ *Pernetti* 202.
‖ *Cook's voy.* ii. 213. *Forfter's voy.* 527. § *Ulloa's voy.* ii. 227.
‡ *Anfon's voy.* 123.

The

The old animals have a tremendous appearance, yet are excessively timid, except at the breeding-feafon, when they feem to lofe their apprehenfions, and are lefs difturbed at the fight of man. At other times, they hurry into the water; or, if awakened out of their fleep by a loud noife, or blows, fall into vaft confufion, tumble down, and tremble in every part, thro' fear.

Thefe animals affociate in families, like the former, but not in fuch great numbers: the males fhew equal jealoufy about their miftreffes, and have bloody combats on their accounts: oft-times there is one of fuperior courage to the reft, and procures by dint of valour a greater number of females than others. They are of a very lethargic nature, fond of wallowing in miry places, and will lie like fwine on one another: they grunt like thofe animals, and will fometimes fnort like horfes in full vigor. They are very inactive on land: to prevent furprize, each herd places a centinel, who gives certain fignals at appearance of danger: during the breeding-feafon, they abftain from food, and before that is elapfed become very lean; at other times they feed on fifh and the fmaller Seals.

487. LEONINE. Beftia Marina, *Kurillis, Kamtfchadalis* et *Ruffis, Kurillico* nomine Siwutfcha dicta. *Nov. Com. Petrop.* ii. 360. Phora Leonina. *Molina Chili.* 262. Sea Lion. *Cook's voy.* ii. 203. *Forfter's voy.* ii. 513. *Pernetti's voy.* 240. *tab.* xvi.

S. with a fhort nofe turning a little up: great head: eyes large: whifkers long and thick, and ftrong enough to ferve for pick-tooths: on the neck and fhoulders of the MALE is a great

mane

Lionine Seal __ N.° 487.

mane of coarfe, long, waving hair, not unlike the fhaggy appearance of a lion: the reft of the body covered with a very
fhort, fmooth, and gloffy coat. The whole color is a deep brown:
thofe of the *Kamtfchatkan* iflands are reddifh ; the females tawny.

The fore feet are like thofe of the *Urfine Seal*, refembling a flat
fin, formed of a black coriaceous fubftance, without the leaft external
appearance of toes, as moft erroneoufly reprefented by *Pernetti:*
the hind feet are very broad, furnifhed with very fmall nails,
with a narrow ftripe of membrane extending far beyond each : tail
very fhort: hind parts vaftly large, fwelling out with the vaft
quantity of fat.

The old males are from ten to fourteen feet long, and of
great circumference about the fhoulders; they weigh from twelve
to fifteen hundred pounds: the females are from fix to eight
feet in length, of a more flender form than the males, and are
quite fmooth.

Penrofe and *Pernetti* afcribe a much greater fize to thofe of the
Falkland ifles. The former fays, that fome of the males are
twenty-fix feet long* ; and the latter affirms that their length is
twenty-five feet, and their girth round the fhoulders from nineteen to twenty †.

They inhabit in vaft numbers *Pinguin* and *Seal* iflands, near
Cape *Defire,* on the coaft of *Patagonia* ‡ ; are found within the
ftraits of *Magellan,* and on *Falkland ifles* : they have not yet been
difcovered in any other part of the fouthern hemifphere, or in
any other place nearer than the fea between *Kamtfchatka* and
America. The inhabitants of *Chili* call them *Thapel lame,* or the
Seal with a *mane.*

Size.

Place.

* *Exped. Falkland Ifles,* 28. † *Voy. Malouines,* 240. ‡ *Narborough,* 31.

They

They live in families feparate from the *Urfine* and other Seals: thefe poffefs the beach neareft to the fea: they have much of the lethargic nature of the former; and, like them, are polygamous: they have from two to thirty females apiece: they have a fierce look; the old ones fnort and roar like enraged bulls; but on the approach of mankind, fly with great precipitation: the females make a noife like calves: the young bleat like lambs.

The old males lie apart, and poffefs fome large ftone, which no other dare approach; if they do, a dreadful combat enfues, and the marks of their rage appear in the deep gafhes on various parts of their bodies. The males frequently go into the water, take a large circuit, land, and carefs their females with great affection; put fnout to fnout as if they were kiffing one another. The females, on feeing their male deftroyed, will fometimes attempt to carry away a cub in their mouth, but oftener defert them through fear.

The food of thefe animals is the leffer Seals, Pinguins, and fifh; but while they are afhore they keep, in the breeding-time, a faft of three or four months; but to keep their ftomachs diftended, will fwallow a number of large ftones, each as big as two fifts.

488. URIGNE.   L'Urigne. Phoca lupina. *Molina Chili.* 255.

S. with the body very thick at the fhoulders, gradually leffening to the hind legs: head like a dog, with the ears clofe cut: nofe fhort and blunt: upper lip cunilineated: fix cutting teeth above; four below: fore foot has four toes inclofed in a membranous fheath, fo as to refemble fins: the hind feet are hid in a

continuation

continuation of the fkin of the back, and have five toes of unequal length, like thofe of the human hand: tail three inches long: the fkin is covered with two forts of hairs, one like that of an ox, the other more hard: the colors various: length from three to eight feet.

Thefe are the *Sea wolves* which navigators fpeak of off the ifland of *Lobos*, near the river *Plata*. They appear in vaft multitudes, meet the fhips, and will even hang by the fides with their paws, and feem to ftare at and admire the crew: then drop off and return to their haunts*. They fwim with incredible fwiftnefs. The natives of *Chili* kill them for the fkins, and for the oil.

PLACE.

* Father *Cattaneo's* firft Letter in the miffions of *Paraguay.* p. 227.

XLIII.
MANATI.

Pinniform fore-legs : hind parts ending in a tail, horizontally flat.   Two teats between the legs.

489. WHALE-
TAILED.

Manati. *Ruſſorum* Morſkuia Korowa.   *De Buffon, Supplem.* vi. 399.
Steller in Nov. Com. Petrop. ii. 294.   Trichecus Borealis. *Gm. Lin.* i. 61. β.
Schreber, ii. 95. *Hiſt. Kamtſchatka,* 132.

T HIS animal in nature ſo nearly approaches the cetaceous
tribe, that it is merely in conformity to the ſyſtematic
writers, that I continue it in this claſs : it ſcarce deſerves the
name of a biped ; what are called feet are little more than pec-
toral fins ; they ſerve only for ſwimming ; they are never uſed to
aſſiſt the animal in walking, or landing ; for it never goes aſhore,
nor ever attempts to climb the rocks, like the *Walrus* and *Seal.*   It
brings forth in the water, and, like the whale, ſuckles its young
in that element : like the whale, it has no voice ; and, like that
animal, has an horizontal broad tail in form of a creſcent, without
even the rudiments of hind feet.

PLACE.

Inhabits the ſeas about *Bering*'s and the other *Aleutian* iſlands,
which intervene between *Kamtſchatka* and *America,* but never appears off *Kamtſchatka,* unleſs blown aſhore by a tempeſt.   Is probably the ſame ſpecies which is found above *Mindanao* * ; but is
certainly that which inhabits near *Rodriguez,* vulgarly called
Diego Reys, an iſland to the eaſt of *Mauritius,* or the iſle of

* *Dampier's voy.* i. 321.

3.   *France,*

France, near which it is likewife found *: Sir *Jofeph Banks* favored me with the fketch of one drawn off this ifland in 1761, by *Ulrike Mole,* of the *Norfolk* man of war. It is likely that this fpecies extends to *New Holland,* where *Dampier* fays he has feen it †.

They live perpetually in the water, and frequent the edges of the fhores; and in calm weather fwim in great droves near the mouths of rivers; in the time of flood they come fo near the land that a perfon may ftroke them with his hand: if hurt, they fwim out to the fea: but prefently return again. They live in families, one near another; each confifts of a male, a female, a half-grown young one, and a very fmall one. The females oblige the young to fwim before them, while the other old ones furround, and, as it were, guard them on all fides. The affection between the male and female is very great; for if fhe is attacked, he will defend her to the utmoft, and if fhe is killed, will follow her corpfe to the very fhore, and fwim for fome days near the place it has been landed at.

They copulate in the fpring, in the fame manner as the human kind, efpecially in calm weather, towards the evening. The female fwims gently about; the male purfues, till, tired with wantoning, fhe flings herfelf on her back, and admits his embraces‡. *Steller* thinks they go with young above a year: it is certain that they bring but one young at a time, which they fuckle by two teats placed between the breaft.

They are vaftly voracious and gluttonous, and feed not only on the *fuci* that grow in the fea, but fuch as are flung on the

* *Voy. de la Caille,* 229. † *Voy.* i. 33.

‡ The *Leonine* and *Urfine* Seals copulate in the fame manner, only, after fporting in the fea for fome time, they come on fhore for that purpofe.

edges of the fhore. When they are filled, they fall afleep on their backs. During their meals, they are fo intent on their food, that any one may go among them and chufe which he likes beft.

Their back and their fides are generally above water; and as their fkin is filled with a fpecies of loufe peculiar to them, numbers of gulls are continually perching on their backs and picking out the infects.

They continue in the *Kamtfchatkan* and *American* feas the whole year; but in winter are very lean, fo that you may count their ribs. They are taken by harpoons faftened to a ftrong cord, and after they are ftruck it requires the united force of thirty men to draw them on fhore. Sometimes, when they are transfixed, they will lay hold of the rocks with their paws, and ftick fo faft as to leave the fkin behind before they can be forced off. When a *Manati* is ftruck, its companions fwim to its affiftance; fome will attempt to overturn the boat, by getting under it; others will prefs down the rope, in order to break it; and others will ftrike at the harpoon with their tails, with a view of getting it out, which they often fucceed in. They have not any voice, but make a noife by hard breathing, like the fnorting of a horfe.

DESCRIPTION.

They are of an enormous fize: fome are 28 feet long, and 8000 lb. in weight; but if the *Mindanao* fpecies is the fame with this, it decreafes greatly in fize as it advances fouthward, for the largeft which *Dampier* faw there, weighed only fix hundred pounds *. The head, in proportion to the bulk of the animal,

* *Dampier*, i. 321. Voyagers are requefted to obferve, whether there are not the two fpecies about this and the other iflands of the *Indian* ocean.

is

is small, oblong, and almost square: the nostrils are filled with short bristles: the gape, or *rictus*, is small: the lips are double: near the junction of the two jaws the mouth is full of white tubular bristles, which serve the same use as the laminæ in whales, to prevent the food running out with the water: the lips are also full of bristles, which serve instead of teeth to cut the strong roots of the sea-plants, which floating ashore are a sign of the vicinity of these animals. In the mouth are no teeth, only two flat white bones, one in each jaw; one above, another below, with undulated surfaces, which serve instead of grinders.

The eyes are extremely small, not larger than those of a sheep: the iris black: it is destitute of ears, having only two orifices, so minute that a quill will scarcely enter them: the tongue is pointed, and small: the neck is thick, and its junction with the head scarce distinguishable; and the last always hangs down. The circumference of the body near the shoulders is twelve feet; about the belly twenty; near the tail only four feet eight: the head thirty-one inches: the neck near seven feet: and from these measurements may be collected the deformity of this animal. Near the shoulders are two feet, or rather fins, which are only two feet two inches long, and have neither fingers nor nails; beneath are concave, and covered with hard bristles: the tail is thick, strong, and horizontal, ending in a stiff black fin, and like the substance of whalebone, and much split in the fore part, and slightly forked; but both ends are of equal lengths, like that of a whale.

The skin is very thick, black, and full of inequalities, like the bark of oak, and so hard as scarcely to be cut with an ax, and has no hair on it: beneath the skin is a thick blubber, which tastes like oil of almonds. The flesh is coarser than beef, and will not

soon

SIZE.

TAIL.

foon putrify. The young ones tafte like veal. The fkin ufed for fhoes, and for covering the fides of boats.

The *Ruffians* call this animal *Morfkaia korowa*, or Sea-cow; and *Kapuftnik*, or Eater of herbs.

490. ROUND-
TAILED.

Adanfon's Senegal. 259. LEV. MUS.

M. with thick lips: eyes as minute as a pea: two very fmall orifices in the place of ears: in each jaw are nine grinding teeth; in all thirty fix: neck fhort, and thicker than the head; the greateft thicknefs of the body is about the fhoulders, from which it grows gradually fmaller to the tail: the tail lies horizontally, is broad, and thickeft in the middle, growing thinner to the edges, and quite round.

The feet are placed at the fhoulders: beneath the fkins are bones for five complete toes, and externally are three or four nails flat and rounded: near the bafe of each leg, in the female, is a fmall teat.

The fkin is very thick and hard, having a few hairs fcattered over it.

SIZE.

The length of the fpecimen in the LEVERIAN MUSEUM is fix feet and a half; the greateft circumference, three feet eight inches; that near the tail, two feet two. This was taken near the *Marigot* of *Kantai*, in the river *Senegal*: they grow to the length of fourteen or fifteen feet: they are very fat, and both fat and lean refemble veal: but the fat adheres to the fkin, in form of blubber: the negroes take them by harpooning, and fell them at the rate of

two

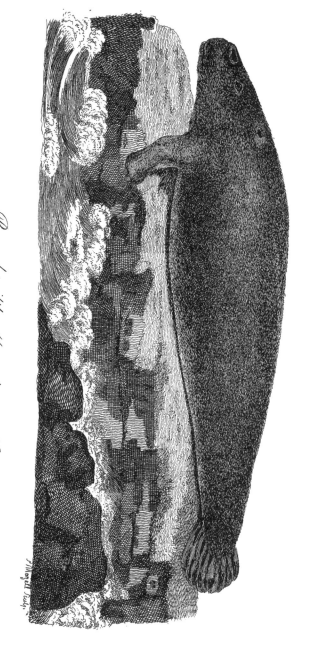

Round-tail'd Manati. — N.º 490.

two long bars of iron apiece. The feafon is only in the months of *December* and *January*. *Manati* are found in moft of the *African* rivers to the fouth of the *Niger*, and poffibly to thofe on the eaftern coaft. The woman-fifh taken off the ifles *Boçicas*, to the fouth of the river *Cuama*, is feemingly of this fpecies, notwithftanding the pious defcriber, Father *Jonanes dos Sanctos*, furnifhes it with four tremendous tufhes *.

De Buffon, xiii. 425. tab. lvii. *Raii fyn.*   Trichechus Manatus. *Lin. fyft.* 49.
quad. 193.   *Schreber,* tab. lxxx.   491. GUIANA.

M. with a head hanging downward; the feet furnifhed with five toes: body almoft to the tail of an uniform thicknefs; near its junction with that part grows fuddenly thin: tail flat, and in form of a *fpatula*; thickeft in the middle, growing thinner towards the edges.

Inhabits the rivers and fea of *Guiana*: it grows to the length of fixteen or eighteen feet: is covered with a dufky fkin with a few hairs †. Thofe meafured by *Dampier* were ten or twelve feet long: their tail twenty inches in length; fourteen in breadth; four or five thick in the middle; two at the edges: the largeft (according to the fame voyager) weighed twelve hundred pounds. But they arrive at far greater magnitude: *Clufius* examined one which was fixteen feet and a half long; and *Gomora* fpeaks of them as fometimes of the length of twenty feet.

* *Purchas.* ii. 1446.   † *Bancroft's Guiana,* 186.

492. MANATI CLUSII.

CLUSIUS, in his Exotics, p. 132, gives a print and description of a *Manati* brought from the *West Indies:* but neither one or the other enables us to define the species. He says that it had short nails and broad feet; and that the tail was broad and shapeless. Till we are better informed we shall suppose it to be the same with the *Guiana. M. de Buffon*, in his Supplement, vi. 396, makes it a distinct species, under the title of *Le grand Lamantia des Antilles.*

493. ORONOKO.

THIS is the species to which *M. de Buffon* has in his Supplement, p. 400, given the name of *Le petit Lamantia de L'Amerique,* and says it is found in the *Oronoko, Oyapoc,* and the rivers of *Amazons.* This pushes its way to the amazing distance we have mentioned. By the description *Gumilla* has given of the tail, it is circular *, and probably must be referred to this species. I do not understand why *M. de Buffon* calls it *Le petit,* for it grows to a vast size. Father *Gumilla* had one taken in a distant lake, near the *Oronoko,* which was so large that twenty-seven men could not draw it out of the water: on cutting it open, he found two young ones, which weighed twenty-five pounds apiece.

We suspect that the *Manati* of the *Amazons,* &c. never visit the sea, but are perpetually resident in the fresh waters.

PLACE.

These animals abound in certain parts of the eastern coasts and rivers of *South America,* about the Bay of *Honduras,* some of the greater *Antilles* †, the rivers of *Oronoque* ‡, and the lakes formed by it; and lastly, in that of the *Amazons,* and the

* *Gumilla,* 54. † *Dampier,* i. 33. ‡ *Gumilla,* ii. 43.

Guallaga,

Guallaga, the *Paftaça*, and moft of the others which fall into that vaft river: they are found even a thoufand leagues from its mouth, and feem to be ftopt from making even an higher advance, only by the great cataract, the *Pongo* of *Borja* *. They fometimes live in the fea, and often near the mouth of fome river, into which they come once or twice in twenty-four hours, for the fake of brouzing on the marine plants which grow within their reach: they altogether delight more in brackifh or fweet water, than in the falt; and in fhallow water near low land, and in places fecure from furges, and where the tides run gently †. It is faid, that at times they frolick and leap to great heights out of the water ‡. Their ufes were very confiderable to the privateers or buccaneers in the time of *Dampier*. Their flefh and fat are white, very fweet and falubrious; and the tail of a young female was particularly efteemed. A fuckling was held to be moft delicious, and eaten roafted, as were great pieces cut out of the belly of the old animals.

The fkin cut out of the belly (for that of the back was too thick) was in great requeft for the purpofe of faftening to the fides of canoes, and forming a place for the infertion of the oars. The thicker part of the fkin, cut frefh into lengths of two or three feet, ferves for whips, and become, when dried, as tough as wood.

In the head, it was pretended that there were certain ftones, or bones of great value, on account of their virtues in curing the gravel and colic ‖.

---

* *Condamine,* 77.   † *Dampier,* i. 34.   ‡ *Gumilla,* ii. 55.
‖ *Clufii Exot.* 233. *Monardus fimp. Med.* 326.

Q q 2   They

They are taken by an harpoon ftuck in the end of a ftaff, which the *Indians* ufe with great dexterity. They go in a fmall canoe with the utmoft filence, as the animal is very quick of hearing. The harpoon is loofe, but faftened to a cord of fome fathoms in length; for as foon as the *Manati* is ftruck, it fwims away with the barb infixed in its body, attended by the canoe, till fpent with pain and fatigue: in fome places the leffer are taken in nets. If a female, which has a young one, is ftruck, fhe takes it under its fins or feet, if not too large, and fhews, even in extremity, the greateft affection for its offspring; which makes an equal return, never forfaking the captured parent, but is always a fure prey to the harpooner [*].

The *Indians* of the *Maragnon,* or the river of *Amazons,* take them by the means of intoxicating herbs, or by fhooting them with thofe poifoned arrows [†], whofe left touch is fatal, yet imparts no degree of venom to the thing ftricken, whofe flefh is eaten with the utmoft fafety [‡].

At the time the waters of the *Oronoque* (which annually overflow the banks) begin to return into the bed of the river, the *Indians* make dams acrofs the mouths of the fhallow lakes formed by the floods, and in that manner take vaft numbers of *Manati,* or *Pexi-buey,* or *Fifh-cows,* as the *Spaniards* call them, together with tortoifes, and variety of fifh [||].

I conclude this account with the extraordinary hiftory of a tame *Manati,* preferved by a certain prince of *Hifpaniola,* at the time of the arrival of the *Spaniards,* in a lake adjoining to his re-

---

[*] *Dampier,* i. 37.　　　[†] *Ulloa,* i. 412. *Gumilla,* ii. 46.　　　[‡] *Condamine's Trav.* 34. *Ph. Tr.* xlvii. 81.　　　[||] *Gumilla,* ii. 43.

fidence.

fidence. It was, on account of its gentle nature, called in the language of the country *Matum*. It would appear as foon as it was called by any of its familiars; for it hated the *Spaniards*, on account of an injury it had received from one of thefe adventurers. The fable of *Arion* was here realifed. It would offer itfelf to the *Indian* favorites, and carry over the lake ten at a time, finging and playing on its back; one youth it was particularly enamoured with, which reminds me of the claffical parallel in the Dolphin of *Hippo*, fo beautifully related by the younger *Pliny*. The fates of the two animals were very different; *Matum* efcaped to its native waters, by means of a violent flood; the *Hipponenfian* fifh fell a facrifice to the poverty of the retired Colonifts*.

M^R. *Steller* faw on the coaft of *America* † another very fingular animal, which he calls a *Sea Ape*; it was five feet long: the head like a dog's: ears fharp and erect: eyes large; on both lips a fort of beard: the form of its body thick and round, thickeft near the head, tapering to the tail, which was bifurcated, the upper lobe the longeft: the body covered with thick hair, grey on the back, red on the belly. *Steller* could difcover neither feet nor paws. It was full of frolick, and played a

494. SEA APE.

---

* See both relations; the firft in *Peter Martyr's Decades of the Indies*, Dec. iii. book 8; the other in lib. ix. epift. 33, of *Pliny*. The elder *Pliny* alfo relates the fame ftory, lib. ix. c. 8.

† The *Beluga*, which I placed here in my former edition, from the mifreprefentation of other writers, is an animal of the cetaceous tribe, called by the Germans, *Witfifh*. See *Pallas Itin.* iii. 84, tab. iv. and *Crantz Greenland*, i. 114. N° 10.

thoufand

thoufand monkey tricks; fometimes fwimming on one fide, fometimes on the other fide of the fhip, looking at it with great
amazement. It would come fo near the fhip, that it might be
touched with a pole; but if any body ftirred, would immediately
retire. It often raifed one-third of its body above the water, and
ftood erect for a confiderable time; then fuddenly darted under the
fhip, and appeared in the fame attitude on the other fide; and
would repeat this for thirty times together. It would frequently
bring up a fea-plant, not unlike the bottle gourd, which it would
tofs about, and catch again in its mouth, playing numberlefs fan-
taftic tricks with it.

D I V.

DIV. IV.

WINGED QUADRUPEDS.

D I V. IV. Winged Quadrupeds:

XLIV. BAT.

With long extended toes to the fore feet, connected by thin broad membranes, extending to the hind legs.

* Without Tails.

495. TERNATE.

Vefpertilio ingens. *Cluf. exot.* 94.
Canis volans ternatanus orientalis. *Seb. Muf.* i. 91. *tab.* lvii.
Vefpertilio Vampyrus. V. ecaudatus, nafo fimplici, membrana inter-femora divifa. *Lin. fyft.* 46.
La Rouffette & la Rougette. *De Buffon,*

x. 55. *tab.* xiv. xvii*. *Schreber,* 185. tab. xliv.
Pteropus rufus aut niger auriculis brevibus acutiufculis. *Briffon quad.* 153, & 154. *No.* 2. *Shaw Spec. Lin.* viii.
Great Bat. *Edw.* 180. *Br. Muf. Afh. Muf.* LEV. MUS.

THE ROUSETTE.

B. with large canine teeth: four cutting teeth above, the fame below: fharp black nofe: large naked ears: the tongue is pointed, terminated by fharp aculeated *papillæ:* exterior toe detached from the membrane: the claw ftrong, and hooked: five toes on the hind feet: talons very crooked, ftrong, and compreffed fideways: no tail: the membrane divided behind quite to the rump: head of a dark ferruginous color: on the neck, fhoulders, and under fide, of a much lighter and brighter red: on the back the hair fhorter, dufky, and fmooth: the membranes of the wings dufky: varies in color; fome entirely of a reddifh brown; others dufky. This now defcribed was one foot long: its extent from tip to tip of the wings four feet; but they are found vaftly larger.

SIZE.

* The Hiftory of thefe bats has been greatly elucidated by M. *De la Nux,* who refided fifty years in the *Ifle de Bourbon,* where they are found. See M. *de Buffon,* Suppl. iii. 253.

This

Ternate Bat ___ N.° 495.

This species is not gregarious, yet they are found in numbers on the same tree, by accidentally meeting there in search of food: they fly by day, and are seen arriving one by one to the spot which furnishes subsistence. If by any accident they are frighted, they will then quit the tree in numbers, and thus fortuitously form a flock. It is different with the other species.

The Rougette*, or BAT, with the same kind of teeth as the other, and the shape of head and body the same: the whole body and head cinereous, mixed with some black; but on the neck is a great bed of lively orange, or red.

THE ROUGETTE.

The size is much less; the extent of wings being little more than two feet.

SIZE.

They are both inhabitants of the same countries, agree in their food, but differ in some of their manners, which I shall distinguish in the following history of them.

These monsters inhabit *Guinea*, *Madagascar*, and all the islands from thence to the remotest in the *Indian* ocean. They are found again in *New Holland*†, the *Friendly islands*, the *New Hebrides*, and *New Caledonia* ‡. The *Rougettes* fly in flocks, and perfectly obscure the air with their numbers: they begin their flight from one neighboring island to another immediately on sun-set, and return in clouds from the time it is light till sun-rise ‖, and lodge during day in hollow trees: both live on fruits; and are so fond of the juice of the *palm*-tree, that they will intoxicate themselves with it till they drop on the ground §.

PLACE.

Notwithstanding the size of their teeth, they are not carnivorous. Mr. *Edwards* relates, that they will dip into the sea for

* LEV. MUS. † *Cook's voy.* iii. 626. ‡ *Forster's obs.* 187. ‖ *Dampier's voy.* i. 381. § *Musæum Hafniæ*, Pars i. Sect. 2, No. 18.

fiſh. I ſuſpect that fact; but it is known that they ſkim the water with wonderful eaſe, perhaps in ſportive moods. They alſo frequent that element to waſh themſelves from any vermin which might adhere to them *. They ſwarm like bees, hanging near one another from the trees in great cluſters †; at leſt five hundred were obſerved hanging, ſome by their fore, others by their hind legs, in a large *Caſuarina*-tree, in one of the *Friendly iſlands*. When ſhot at, they flew from the boughs very heavily, uttering a ſhrill piping note; others again, arrived at intervals from remote places to the tree‡. In *New Caledonia*, the natives uſe their hair in ropes, and in the taſſels of their clubs, interweaving it with the threads of the *Cyperus ſquarroſus*. The *Indians* eat them, and declare the fleſh to be very good: they grow exceſſively fat at certain times of the year. The *French*, who live in the *Iſle de Bourbon*, boil them in their *bouillon*, to give it a reliſh ‖. The *Negroes* have them in abhorrence §. Many of the *Rouſſettes* are of an enormous ſize: *Beeckman* ** meaſured one, whoſe extent from tip to tip of the wing was five feet four inches; and *Dampier* †† another, which extended further than he could reach with ſtretched-out arms. Their bodies are from the ſize of a pullet to that of a dove: while eating, they make a great noiſe: their ſmell rank; their bite, reſiſtance, and fierceneſs great when taken.

They bring but one young at a time.

The antients had ſome knowledge of theſe animals. *Herodotus* ‡

---

* *Forſter's obſ.* 188.
‡ *Forſter's voy.* i. 450.
** *Voy. to Borneo*, 39.
πϱοσιιχιλα. *Lib.* iii.

† *Argenſola Philip. iſles*, 158. *Des Marchais*, ii. 261.
‖ *Voy. de la Caille*, 233.   § *Des Marchais*, ibid.
†† i. 381.   ‡ Θηϱια πλεϱωτα, τησι νυκτεϱισι

mentions

mentions certain winged wild beasts, like bats, that molested the *Arabs*, who collected the *Caffia*, to such a degree that they were obliged to cover their bodies and faces, all but their eyes, with skins. It is very probable, as M. *de Buffon* remarks, it was from such relations the Poets formed their fictions of *Harpies*.

Linnæus gives this species the title of *Vampyre*, conjecturing it to be the kind which draws blood from people in their sleep. M. *de Buffon* denies it, ascribing that faculty to a species only found in *S. America*: but there is reason to imagine, that this thirst after blood is not confined to the bats of one continent, nor to one species; for *Bontius* and *Nieuhoff* inform us, that they of *Java* * seldom fail attacking those who lie with their feet uncovered, whenever they can get access; and *Gumilla* †, after mentioning a greater and lesser species, found on the banks of the *Orenoque*, declares them to be equally greedy after human blood. Persons thus attacked, have been known to be near passing from a sound sleep into eternity. The Bat is so dextrous a bleeder as to insinuate its aculeated tongue into a vein without being perceived, and then suck the blood till it is satiated; all the while fanning with its wings, and agitating the air, in that hot climate, in so pleasing a manner, as to fling the sufferer into a still sounder sleep ‡. It is therefore very unsafe to rest either in the open air, or to leave open any entrance to these dangerous animals: but they do not confine themselves to human blood; for M. *Con*-

---

* *Bontius India*, 70. *Nieuhoff*, 255. These writers say that this kind is as big as a pigeon. I suspect that the species just described is common to *India* and *S. America*; Mr. *Greenwood*, painter, long resident at *Surinam*, informing me that there is in that colony a *fox-colored* bat, whose extent of wings is above four feet.

† *Hist. Orenoque*, iii. 100.   ‡ *Ulloa's voy.* i. 61.

damine

damine * fays, that in certain parts of *America* they have deftroyed all the great cattle introduced there by the miffionaries.

β. LESSER. B. with head like a grehound: large teeth like the former: ears long, broad, and naked: whole body covered with foft fhort hair of a ftraw-color: fhaped like the other in all refpects: length, eight inches three quarters; extent, two feet two inches. Place unknown to the gentleman who favored me with it. LEV. MUS.

496. SPECTRE.

Andira-guacu, vefpertilio cornutus. *Pifo Brafil.* 190. *Marcgrave Brafil.* 213. Canis volans maxima aurita fæm. ex Nov. Hifpania. *Seb. Muf.* i. *tab.* lvii. Vefpertilio fpectrum: V. ecaudatus, nafo infundibuliformi lanceolato. *Lin. fyft.*

46. *Klein quad.* 62. Pteropus auriculis longis, patulis, nafo membrana antrorfum inflexa aucto. *Briffon quad.* 154. Le Vampire. *De Buffon,* x. 55. *Schreber,* 192. *tab.* xlv.

B. with a long nofe: large teeth: long, broad, and upright ears: at the end of the nofe a long conic erect membrane, bending at the end, and flexible: hair on the body cinereous, and pretty long: wings full of ramified fibres: the membrane extends from hind leg to hind leg: no tail; but from the rump extend three tendons, terminating at the edge of the membrane. By

SIZE.

Seba's figure, the extent of the wings is two feet two inches; from the end of the nofe to the rump feven inches and a half.

PLACE.

Inhabits *South America:* lives in the palm-trees: grows very fat: called *Vampyre* by M. *de Buffon,* who fuppofes it to be the fpecies that fucks human blood: but neither *Pifo,* or any other writers who mention the fact, give the left defcription of the kind.

* *Voy. S. America,* 85.

Vefpertilio

Vespertilio Americanus vulgaris. *Seb.*
  Muf. i. *tab.* lv. *fig.* 2.
Vespertilio ¡erspicillatus. V. ecaudatus,
  naso foliato acuminato. *Lin.fyft.* 47.
V. murini coloris pedibus anticis tetra-
  dactylis, posticis pentadactylis. *Bris-*

fon quad. 161.
La chauve souris fer de Lance. *De Buffon,*
  xiii. ≈26. *tab.* xxxiii. *Supplem.* vii. 292
  tab. lxxiv. *Schreber,* 194. tab. xlvi.
  B.

B. with large pointed ears : an erect membrane at the end of
  the nose, in form of the head of an antient javelin, having
on each side two upright processes : no tail : fur cinereous : size
of a common bat.

Inhabits the warm parts of *America.*

The bat described by Mr. *Schreber,* p. 193. tab. xlvi. A. under
the title of *La Chauve souris pelle,* has so much resemblance, that I
place it here as a variety of the former : the nasal membrane being nearly of the same form; the color differs, the fur being ferruginous.

Vespertilio, rostro appendice auriculæ
  forma donata *Sloane Jam.* ii. 330.
Small bat. *Edw.* 201. *fig.* 1.
La Feuille. *De Buffon,* xiii. 227.

Vespertilio soricinus. *Pallas Miscel.* 48.
  · *tab.* v.* *Schreber,* 195. tab. xlvii.
  LEV. MUS.

B. with small rounded ears : membrane on the nose of the form
  of an ovated leaf : no tail : a web between the hind legs :
fur of a mouse-color, tinged with red : size of the last.

* This seems to be one of the blood-sucking species, the tongue being furnished
with aculeated *papillæ,* and is twice the length of the nose ; so is well adapted for
that purpose.

Inhabits

PLACE.

Inhabits *Jamaica*, *Surinam*, and *Senegal*: in the firſt lives in caves in woods, which are found full of its dung, productive of ſalt-petre : feeds on the prickly pear.

499. CORDATED.

Glis volans Ternatanus. *Seb. Muſ.* i *tab.* lvi. *fig.* 1. *Schreber*, 191. tab. xlviii.   Veſpertilio ſpaſina. V. ecaudatus naſo fo- liato obcordato. *Lin. ſyſt.* 47.

B. with very broad and long ears : at the end of the noſe a heart-ſhaped membrane : no tail : a web between the hind legs : color of the face a very light red; that of the body ſtill paler.

PLACE.

Inhabits *Ceylon*, and the iſle of *Ternate*, one of the *Moluccas*.

* * With tails.

500. PERUVIAN.

Chauve-ſouris de la Vallée d'Ylo. *Feuillée*   tab. lx. *obſ. Peru*, 1714. *p.* 623. *Schreber* 196.   Veſpertilio Leporinus. *Gm. Lin.* 47.

B. with a head like a pug-dog : large ſtrait ears, ſharp at the ends and pointing forwards : two canine teeth, and two ſmall cutting teeth between each, in each jaw : tail encloſed in the membrane which joins to each hind leg, and is alſo ſupported by two long cartilaginous ligaments involved in the membrane : color of the fur iron-grey; but erroneouſly colored in the print, of a ſtraw color : body equal to that of a middle-ſized rat : extent of wings two feet five inches.

SIZE.

β. With

β. With a large head and hanging lips, like the chops of a maſtiff: noſe bilobated: upper lip divided: ſtrait, long, and narrow ears, ſharp-pointed: teeth like the former: tail ſhort; a few joints of it ſtand out of the membrane, which extends far beyond it; is angular, and ends in a point: claws on the hind feet large, hooked, and compreſſed ſideways: membranes of the wings duſky, very thin: fur on the head and back brown; on the belly, cinereous. Length, from the noſe to the end of the membrane, above five inches; extent of wings, twenty.

Inhabits *Peru* and the *Moſquito* ſhore: the laſt was given me by *John Ellis*, Eſq; F. R. S. It differed from the former in ſize, being leſs; in all other reſpects agreed.

Linnæus, carried away by love of ſyſtem, places this, on account of its having only two cutting teeth in each jaw, among the *Glires*, next to the ſquirrels, under the name of *Noctilio Americanus*. But ſuch is the variety in the numbers and diſpoſition of the teeth in the animals of this genus, that he might form almoſt as many genera out of it as there are ſpecies. But as the Bats have other ſuch ſtriking characters, it is unneceſſary to have recourſe to the more latent marks to form its definition. The ſame may be ſaid of ſeveral other animals.

SIZE.

PLACE.

Autre Chauve ſouris. *De Buffon*, x. 84, 87. *tab.* xix. *fig.* 1, 2. *Schreber*, 207. tab. xlix. LEV. MUS.

501. BULL-DOG.

B. with broad round ears, the edges touching each other in front: noſe thick: lips pendulous: upper part of the body of a deep aſh-color; the lower paler: tail long; the five laſt

joints

SIZE.　　joints quite difengaged from the membrane. Length above two inches; extent nine and a half.

PLACE.　　Inhabits the *Weft Indies.*

502. SENEGAL.　　Chauve-fouris etrangere. *De Buffon,* x. 　　LEV. MUS.
　　　　82. *tab.* xvii. *Schreber,* 206. tab. lviii.　Vefpertilio nigrita. *Gm. Lin.* 49.

B. with a long head: nofe a little pointed: ears fhort, and pointed: head and body a tawny brown mixed with afh-color: belly paler: two laft joints of the tail extend beyond the

SIZE.　　membrane. Length from nofe to rump, above four inches; extent 21.

PLACE.　　Inhabits *Senegal.*

503. POUCH.　　La Chauve-fouris a bourfe. *Schreber,* 209. tab. lvii.

WITH the nofe fomewhat produced: the end thickeft, and befet with fine whifkers: the chin divided by a fulcus: ears long, rounded at their ends: on each wing, near the fecond joint, is a fmall purfe, or pouch: the tail is only partly involved in the membrane; the end hanging out: color of the body a cinereous brown: the belly paler.

SIZE.　　Length an inch and a half.
PLACE.　　Inhabits *Surinam.*

Autre

Autre Chauve-fouris *de la Guyanne. De Buffon, Supplem.* vii. 214. tab. lxxv.

B. with large pendulous ears, pointed at the ends: nose obtufe at the end: tail long, included in the membrane, and ending with a hook: color above, a deep chefnut; lighter on the belly, and cinereous on the fides: length three inches and four lines: extent of wings fifteen inches.

Inhabits *Guiana.*

Autre Chauve-fouris. *De Buffon,* x. 92. *tab.* xx. *fig.* 3. *Schreber,* 204. *tab.* lvi. LEV. MUS.

B. with the noftrils open for a great way up the nofe: hair on the forehead and under the chin very long: ears long and narrow: upper part of the head and body of a reddifh brown; the lower of a dirty white tinged with yellow: tail included in a membrane very full of nerves. A fmall fpecies.

B. with a head fhaped like that of a moufe: top of the nofe a little bifid: ears fhort, broad, and rounded: no cutting teeth; two canine in each jaw: tail very long, inclofed in the membrane, which is of a conic fhape: head, body, and the whole upper fide of the membrane which inclofes the tail, covered with long very foft hair of a bright tawny color, lighteft on the head and beginning of the back; the belly paler: at the bafe of each wing a white fpot: wings thin, naked, and

SIZE.   dufky: bones of the hind legs very flender. Length, from nofe to tail two inches and a half; tail one inch eight-tenths; extent of wings ten and a half.

PLACE.   Inhabits *North America.* Communicated by Mr. *Afhton Black-burne* *. It is alfo found in *New Zeland* †. Mr. *Schreber* de-fcribes it from me, in p. 212. LEV. MUS.

507. STRIPED.   Autre Chauve-fouris. *De Buffon,* x. 92. *tab.* xx. fig. 3. *Zooph. Gronov. No.* 25. *Schreber,* 205. tab. xlix.

B. with a fmall fhort nofe; ears fhort, broad, and pointing forward: body brown: wings ftriped with black, and

SIZE.   fometimes with tawny and brown. Length, from nofe to the end of the tail, two inches: varies in color; the upper part of the body being fometimes of a clear reddifh brown, the lower whitifh.

PLACE.   Inhabits *Ceylon*; called there, *Kiriwoula* ‡. I may add to this little fpecies of Bat, the mention of a minute kind feen and heard in myriads of numbers in the ifle of *Tanna,* one of the *New He-brides,* but which efcaped every attempt of our voyagers to ob-tain a near examination ‖.

---

* The Rev. M. *Clayton* mentions another fpecies of *North American* Bat; large, with great ears, and long ftraggling hairs. *Phil. Tranf. abridg.* iii. 594.

† *Forfter's obferv.* 189.   ‡ *Pallas Mifcel.* 49.   ‖ *Forfter's obf.* 188.

Vefpertilio

2

1

1. Lesser Ternate. Bat p. 308
2. New York — N.° 506

S. Mazell Sculp.

Vefpertilio Cephalotes. *Pallas Spicil. Zool. fafc.* iii. 10. *tab.* i. *Schreber,* 208. *tab.* lxi. Lev. Mus.

508. MOLUCCA.

B. with a large head: thick nofe: fmall ears: tubular noftrils, terminating outwards in form of a fcrew: upper lip divided: tongue covered with papillæ and minute fpines: claw, or thumb, joined to the wing by a membrane: firft ray of the wing terminated by a claw: end of the tail reaches beyond the membrane: color of the head and back greyifh afh-color; that in the LEVERIAN MUSEUM of a fine ftraw-color: the belly dull white. Length, from nofe to rump, three inches three quarters; extent of wings about fifteen.

SIZE.

Inhabits the *Molucca* ifles. Defcribed firft by that very able naturalift Doctor *Pallas.*

PLACE.

Vefpertilio Lepturus. *Schreber,* tab. lvii. *Gm. Lin.* 50.

509. SLENDER-TAILED.

B. with tubular noftrils: long erect ears: color dufky above, cinereous beneath.

Inhabits *Surinam.*

PLACE.

Vefpertilio Lafiurus. *Schreber, tab.* lxii. *Gm. Lin.* 50.

510. ROUGH-TAIL'D.

B. with upright fmall ears: tail broad at the bafe, terminating in a point thickly covered with hair: color a reddifh brown: a fmall fpecies.

Place unknown.

S s 2

Vefpertilio

511. LASCOPTE-
RUS.

Vefpertilio Lafcopterus. *Schreber*, tab. lviii. B. *Gm. Lin.* 50.

B. with a moft prominent rounded forehead : fhort nofe : color a
bright ruft : upper part of the wings of a paler ruft : ends and
lower parts of the wings black.   By Mr. *Schreber's* figure it feems a
large fpecies.

Place unknown.

512. HORSE-SHOE.

La Chauve-fouris fer a Cheval. *De Buffon*, viii. 131, 132. *tab.* xvii. xx. *Schreber*,
210. tab. lxii. *Br. Zool.* i. 129.

B. with a membrane at the end of the nofe in form of a horfe-
fhoe : ears large, broad at their bafe, and fharp-pointed, inclining backward : wants the little or internal ear : color of the
upper part of the body deep cinereous ; of the lower, whitifh.
There is a greater and leffer variety ; the greater is above three

SIZE.

inches and a half long from the nofe to the tip of the tail : its
extent above fourteen.   This and all the following have the tail
inclofed in the membrane.

PLACE.

Inhabits *Burgundy* ; and has lately been difcovered in *Kent*, by
Mr. *Latham*, of *Dartford* ; found alfo about the *Cafpian* fea.   The
long-eared Bat, No 519, has alfo been obferved there, and at *Peterfburg*.   This and the four next were firft difcovered by M. *de Buffon*, whofe names I retain.

La Noctule. *De Buffon,* viii. 128. *tab.* xviii. *Schreber,* 200. tab. lii.
Great Bat. *Br. Zool. illustr. tab.* ciii. *Br. Zool.* i. 128.

B. with the nose slightly bilobated: ears small and rounded: on the chin a minute *verruca:* hair of a reddish ash-color. Length to the rump two inches eight-tenths; tail one seven-tenth; extent of wings thirteen inches.

Inhabits *Great Britain* and *France*; very common in the open deserts of *Ruffia,* wherever they can find shelter in caverns: flies high in search of food, not skimming near the ground. A gentleman informed me of the following fact, relating to those animals, which he was witness to:—that he saw taken under the eaves of *Queen's College, Cambridge,* in one night, one hundred and eighty-five; the second night sixty-three; the third night two; and that each that was measured had fifteen inches extent of wings *.

La Serotine. *De Buffon,* viii. 129, tab. xviii. *Schreber,* 201. tab. liii.

B. with a longish nose: ears short, but broad at the base: hair on the upper part of the body brown, mixed with ferruginous; the belly of a paler color. Length from nose to rump, two inches and a half: no tail.

* No notice was taken of the species; but, by the size, it could be neither of the common kinds. I never saw but one specimen of the *Noctule,* which was caught during winter in *Flintshire.*

Inhabits

PLACE.   Inhabits *France*; found in caverns of rocks upon the river *Argun*, beyond lake *Baikal*; but as yet not difcovered in any other part of the vaft *Ruffian* dominions.

515. GREATER   La Grande Serotine de la *Guyanne*. *De Buffon Supplem.* vii. 289. *tab.* lxiii.
SEROTINE.

B. with a very long, ftrait and ftrong nofe, floping down at the end : ears long, erect, dilated towards the bottom, rounded at the end : color of the upper parts of a reddifh chefnut ; fides a clear yellow, reft of a dirty white.   Length five inches eight lines: extent of wings two feet : no tail.

PLACE.   Inhabits *Guiana*: affembles in vaft numbers in open places, particularly meadows ; and flies in company with the goat-fuckers, and both together, in fuch numbers as to darken the air.

516. PIPIST-   La Pipiftrelle. *De Buffon*, viii. 129. tab. xix. *fig.* 2. *Schreber*, 202. tab. liv.
RELLE.

B. with a fmall nofe : the upper lip fwelling out a little on each fide: the ears broad : the forehead covered with long hair : color of the upper part of the body a yellowifh brown ; the lower part dufky ; the lips yellow.   The leaft of Bats ; not an inch and a quarter long to the rump: extent of wings fix and a half.

SIZE.

PLACE.   Inhabits *France*: common in the rocky and mountanous parts of *Ruffia* and *Siberia*.

La

La Barbaftelle. *De Buffon*, viii. 130. *tab.* xix. *fig.* 1. *Schreber*, 203. tab. lv.

B· with a funk forehead: long and broad ears; the lower part of the inner fides touching each other; and conceal the face and head when looked at in front: the nofe fhort; the end flatted: cheeks full: the upper part of the body of a dufky brown; the lower, afh-colored and brown. Its length to the rump about two inches; its extent ten and a half.

SIZE.

Inhabits *France*.

PLACE.

Νυκτερις. *Arift. hift. an. lib.* i. *c.* 5.
Vefpertilio. *Plinii lib.* x. *c.* 61. *Gefner quad.* 766. *Agricola Anim. Subter.* 483.
Bat, Flitter-moufe. *Raii fyn. quad.* 243.
Rear-moufe. *Charlton Ex.* 80.
Vefpertilio major. Speck-maus, Fledermaus. *Klein quad.* 61.
Vefpertilio murinus. V. caudatus nafo

oreque fimplici, auribus capite minoribus. *Lin. fyft.* 47.
Laderlap, Fladermus. *Faun. fuec. No.* 2.
La grande Chauve-fouris de notre pais. *Briffon quad.* 158. *De Buffon*, viii. 113. *tab.* xvi.
Short-eared Bat. *Br. Zool.* i. 130. *Edw.* 201. *Schreber*, 199. tab. li. LEV. MUS.

B· with fhort ears: moufe-colored fur tinged with red. Length two inches and a half; extent of wings nine.

SIZE.

Inhabits *Europe*: the moft common fpecies in *England*.

PLACE.

Souris

519. LONG-EAR-ED.

Souris Chauve, Rattepenade. *Belon oyf.* 147.
Vefpertilio auritus. V. nafo oreque fim-
plici, auriculis duplicatis, capite ma-
joribus. *Lin. fyft.* 47. *Faun. fuec. No. 3.*
Klein quad. 61.
La petite Chauve-fouris de notre pais.

Briffon quad. 160. *Shaw fpec. Lin.* vii.
L'Oreillar. *De Buffon*, viii. 118. *tab.*
xvii. *Schreber*, 197. tab. l.
Long-eared Bat. *Edw.* 201. *Br. Zool.* i.
129. *Br. Zool. illuftr. tab.* ciii. LEV.
MUS.

SIZE.

PLACE.

B. with ears above an inch long, thin, and almoft pellucid: body and tail only one inch three quarters long. This and all other Bats, except the *Ternate* and the *Horfe-fhoe*, have a leffer or internal ear, ferving as a valve to clofe the greater when the animal is afleep.

Inhabits *Europe*, and is found in *Great Britain*. Bats appear abroad in this country early in the fpring; fometimes are tempted by a warm day to fally out in winter; fly in the evenings; live on moths and other nocturnal infects; fkim along the water in queft of gnats; fly by jerks, not with the regular motion of birds, for which the antients miftake them; frequent glades and fhady places; will go into larders, and gnaw any meat they find: bring two young at a time, which they fuckle at their breaft: retire at the end of fummer into caves, the eaves of houfes, and into ruined buildings, in vaft multitudes, where they gene-rally remain torpid, fufpended by the hind legs, enveloped in their wings: are the prey of owls: their voice weak. *Ovid* takes notice both of that and the origin of the *Latin* name.

Minimam pro corpore vocem
Emittunt; peraguntque leves ftridore querelas.
Tectaque, non fylvas celebrant: lucemque perofæ
Nocte volant: feroque trahunt a vefpere nomen.

6
ADDITIONS.

ADDITIONS.

A PROOF of their being prolific was produced by Mr. *Tullo*, in the parish of *Newtyle*, in the shire of *Forfar*, about twenty years ago, when a she-mule, which he turned to a horse, brought a foal which much resembled the female parent. But as there is a superstition in *Scotland* about these productions, the foal was put to death, being considered as a monster.

MULES.
page 8.

SUMATRAN ANTELOPE,

ANTELOPE.
p. 104.

As communicated by Doctor *Shaw*.

Cambing ootan, or Goat of the woods, *Marsden's Sumatra*, 93.

SIZE of a common goat, but stands considerably higher on its legs: color an uniform black, but each hair when narrowly examined is grey towards the base: on the top of the neck just above the shoulders a patch of whitish, bristly, long strait hair, much stronger than the rest, and having somewhat the appearance of a partial mane: on each side of the lower jaw a longitudinal patch of yellowish white: ears moderate, marked internally with three obscure longitudinal bands of white, as in some of the antelopes: horns six inches long, bending slightly backwards, sharp-pointed, black and annulated near half their length with

VOL. II.　　　　　　T t　　　　　　　　prominent

prominent rings: tail about the length of horns, and fharpifh: hoofs rather fmall and black: hair on the whole animal rather harfh, and not lighter-colored below, or on the belly, than above.

In its difpofition it is wild and fierce, and is faid by the natives to be remarkably fwift: we are obliged to the author of the elegant hiftory of *Sumatra* for the difcovery of this animal.

MONKEY.
p. 226.

PROBOSCIS MONKEY.

La Guenon a long nez. *De Buffon Supplem.* vii. 53. tab. xi. xii.

M. with the nofe projecting very far beyond the mouth, like the human, but divided in the middle by a fhallow furrow: in the profile it exactly refembles a long probofcis, and makes a ridiculous appearance: the forehead hangs far over the bafe of the nofe: the face is hooked, of a brown color, marked with blue and red: the head covered with thick hair of a chefnut brown: the ears broad, thin, and naked, hid in the fur: the body is large, cloathed with hair of a brown chefnut color; orange on the breaft: round the throat, neck, and fhoulders, the hair is longer than that on the reft of the body, and forms a fort of fhort cloak, of a color contrafting that of the face: the legs are covered with fhort tawny hair:

SIZE. the length from the tip of the nofe to the bafe of the tail is two feet: of the tail, above two feet.

PLACE. Inhabits the *Eaft Indies*; but the particular part is not mentioned.

THE

Proboscis Monkey

Proboscis Monkey.

Heart-marked Maucauco.

P. Mazell Sculp.

THE HEART-MARK'D MAUCAUCO.

I HAVE totally forgotten the friend who obliged me with the drawing of this animal, and the place it came from; but probably from *Madagascar*, or the neighboring isles, the seat of most of the congenerous species.

All the upper parts of the body are of a deep cinereous brown: the face marked with large white heart-shaped spots: the broader part extends between the ears; the point reaches almost to the nose: the belly, legs, and feet, are white. I am at a loss for the size; but possibly the gentleman from whom I received the drawing may reveal himself, and communicate the wanted particulars.

ONE which was examined at the *Cape of Good Hope*, by Captain *Blanket*, had ears like those of a lurcher, but larger, and more on the top of the head. It could turn them on all sides with great facility: feet flatter than those of other dogs. It could not bark or howl, but only cried: was very fierce, and mastered the tame dogs it was with, though it was only a young one.

SLENDER TOED WEESEL.

W with short rounded ears: fur soft and fine, grizzled minutely with black and rufous: toes very long and slender; five in number; each lobated at the bottom of the first

T t 2   joint:

joint: claws fmall: the upper part of the toes and part of the legs covered with fhort velvet-like down.

Length from the tip of the nofe to the bafe of the tail feven inches: tail about the fame length; bufhy or covered with long hairs of the fame color with the rat.

PLACE.   A native of *Cochin China.*

THE ERMINED WEESEL.

W with ears fhort, round and naked: within of a fine pink color: tip of the nofe black: head white and plain; the reft of the body and tail white: the firft fpotted with ermine-like black fpots, difpofed in rows from neck to tail, on the fides as well as back: the tail annulated with black: the hairs on all parts of the tail fhort, only the end is tufted with black.

The legs remarkably ftrong, and thick covered to the very claws with long bright ferruginous hairs: claws fharp and white: length of the head three inches and a half; of the neck and body from head to tail fixteen inches and a half; of the tail eleven and a half.

This elegant animal is likewife a native of *Cochin China,* and with the former, communicated to us by the friendfhip of Lieut. Col. *Davies,* of the artillery.

INDEX.

Slender toed Weesel

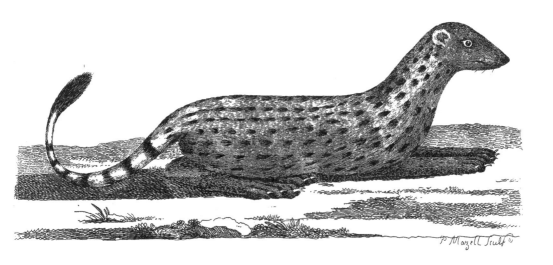

Ermined Weesel

I N D E X*.

A

		Vol.	Page
ANT-EATER, or Ant-Bear		II.	256
ANTELOPES, their general history	—	I.	68
———— Species of	—	I.	70
APES, their general history	—	I.	178
——— Sea	———	—	II. 301
ARMADILLO	———	—	II. 246
Afs	———	—	I. 8
——— Wild	———	—	I. 8
Axis	———	—	I. 117
——— Greater	———	—	I. 118

B

Baboons	———	—	I. 188
Baby-rouffa	———	—	I. 148
BADGER	———	—	II. 14
BATS	———	—	II. 304
BEAR	———	—	II. 1
——— Polar	———	—	II. 5
BEAVER	———	—	II. 114
——— Its wondrous œconomy	—	II. 115
——— Sea, vide Sea Otter.	—	II. 83
Beaver-Eater	———	—	II. 9
Beluga	———	—	II. 301
Bezoar	———	—	I. 58
Bifon, Scottifh	—	—	I. 17
Buck	———	—	I. 113
Buffalo, Indian	—	—	I. 28
——— When introduced into Europe	—	I. 29
——— American	—	I. 23
——— Dwarf, or Anoa	— I. 30, 36
——— Naked, or Bonafus	I. 30
——— Ceylon	—	I. 31
Bull	———	—	I. 16

		Vol.	Page
Bull-Dog	———	—	I. 242

C

CAMEL, Arabian	—	I. 129
——— Bactrian	—	I. 132
——— Peruvian, or Llama	—	I. 133
——— The only native beaft of burden in America	—	I. 134
——— Vicunna	—	I. 136
——— Paicos	———	—	I. 137
——— Guanaco	———	—	I. 138
——— Chilihucque	—	I. 138
Camelopard	———	—	I. 65
Cafioreum	———	—	II. 118
CAT, Common	———	—	I. 295
——— Wild	———	—	I. 296
——— Tiger	———	—	I. 277
——— Mountain	—	—	I. 300
——— Civet	———	—	II. 70
——— Angora	———	—	I. 296
CAVY, various fpecies of	—	II. 88
Chamois	———	—	I. 72
Chimpanzee	———	—	I. 180
Civet	———	—	II. 70

D

DEER,	———	—	I. 105
——— Elk, or Moofe	—	I. 105
——— Rein	———	—	I. 111
——— Fallow	———	—	I. 113
——— Mexican	———	—	I. 122
——— Porcine	———	—	I. 119
——— Grey	———	—	I. 123
——— Virginian	—	I. 116
——— Red, Stag, or Hart	—	I. 114

* In this Index very few of the Species are enumerated, that having been amply done under the INDEX of GENERA; to which the Reader is referred, the Genera being here printed in capitals for that purpofe, under which he will find all the Species belonging to each.

I N D E X.

Vol. Page

DEER, Axis ——— I. 117, 118
——— Rib-faced — I. 119
——— Roe — I. 120
——— Tail-lefs — I. 121
DOGS, the different varieties — I. 235
——— Wild — I. 236
Dormoufe, Common — II. 157
Dromedary — I. 129
Dfhikketaei — I. 4

E

ELEPHANT — I. 165
——— Teeth — I. 172
——— American — I. 174
Elk — I. 105
Ermine — II. 35

F

Ferret — II. 40
Fifher — II. 50
Fitchet — II. 37
Flitter-Moufe — II. 319
Foffane — II. 75
Foumart — II. 33
Fox — I. 251
— Crofs — I. 251
— Brant — I. 252
— Corfak — I. 253
— Arctic — I. 255
— Grey — I. 259
— Silvery — I. 260

G

Gazelle, vide Antelope — I. 89
Genet — II. 74
GIRAFFE — I. 65
Glutton — II. 10
Gnou — I. 70
GOAT, Wild, or Ibex — I. 55
——— Domeftic — I. 60
——— Angora — I. 61
——— Syrian, or long-eared I. 63
——— African — I. 64
——— Caucafan — I. 57
——— Whidaw — I. 63
——— Capricorn — I. 64

Vol. Page

GOAT, Pudu — I. 64
Greyhound — I. 241
Guanaco — I. 138

H

Hamfter — II. 206
HARE — II. 98
——— Alpine — II. 107
——— Baikal — II. 104
Hart — I. 114
HEDGE-HOG — II. 234
HIPPOPOTAME — I. 157
Hog — I. 140
HORSE — I. 1
——— Wild — I. 2
——— Sea, vide Hippopotame.
Hound — I. 239
HYÆNA — I. 270
——— Spotted — I. 272

I

Jackal — I. 272
Ibex — I. 55
Ichneumon, deftroyer of ferpents II. 54
Jerboa — II. 164

K

Kangaru — II. 29
Karagan — I. 252

L

Lamantia — II. 298
Leming — II. 198
Leopard — I. 282
Lion — I. 274
Lizard, Scaly — II. 252
Llama — I. 133
Lynx — I. 301
——— Bay — I. 303
——— Cafpian — I. 304

M

Mammouth's bones — I. 172
Man of the Wood — I. 191
Manipouris — I. 163
MANATI — II. 292

Mandril

INDEX.

	Vol. Page
Mandrill	— I. 190
Manis	— II. 252
MARMOTS	— II. 128
Martin	— II. 41
—— Pine	— II. 42
MAUCAUCOS	— I. 227
Minx	— II. 81
Mococo	— I. 230
Mole-Rats	— II. 214
MOLES	— II. 229
Mongooz	— I. 229
MONKIES	— I. 199
Moose	— I. 105
MORSE	— II. 266
Mouse	— II. 184
MULE, Wild	— I. 4
Mule	— I. 8
Musimon	— I. 44
MUSK. Animal	— I. 124
—— Rat	— II. 221

N

Norway Rat	— II. 178

O

Once	— I. 285
OPOSSUM	— II. 18
Orang Outang	— I. 180
OTTER	— II. 77
Otter, Sea	— II. 83
Ox	— I. 16
—— Great Indian	I. 20, 21
—— Abyssinian	— I. 21
—— Madagascar	— I. 21
—— Tinian	— I. 21
—— Lant or Dant	— I. 21
—— Holstein and Jutland	— I. 21
—— Podolian and Hungarian	— I. 22
—— Grunting	— I. 24
—— Musk	— I. 31
—— Cape	— I. 32
—— American	— I. 23

P

Pacos	— I. 137

	Vol. Page
Panther	— I. 280
Pecary	— I. 147
Pekan	— II. 51
Pig, Guinea	— II. 90
Pole-cat	— II. 37
—— American	— II. 64
PORCUPINE	— II. 122
—— incapable of darting its quills	— II. 123
Potto	— II. 59
Puma	— I. 289
Pygmies, what	— I. 183

Q

Quagga	— I. 14
Quick-hatch	— II. 8
Quojas Morrou	— I. 180

R

Rabbet	— II. 103
Raccoon	— II. 12
RAT	— II. 172
—— Norway	— II. 178
—— Water	— II. 182
—— Musk	— II. 221
Ratel	— II. 66
Rein Deer	— I. 111
RHINOCEROS	— I. 150
River Hog	— II. 88
Roebuck	— I. 120

S

Sable	— II. 43
Schakal	— I. 261
Sea Bear	— II 281
—— Ape	— II. 301
—— Calf	— II. 270
—— Cow	— II. 266
—— Horse	— II 266
—— Lion	— II. 286
SEAL	— II. 270
SHEEP	— I. 37
—— Cretan	— I. 38
—— Hornless	— I. 39
—— Many-horned	— I. 39
African	

INDEX.

	Vol. Page
SHEEP, African	— I. 40
———— Broad-tailed	— I. 41
———— Sibirian	— I. 44
———— Corsican	— I. 45
———— Bearded	— I. 52
SHREW Mouse	— II. 221
Siyah Ghush	— I. 305
Skunk	— II. 65
SLOTH	— II. 240
SQUIRRELS	— II. 138
Stag	— I. 114
Stoat	— II. 35
Strepsiceros	I. 38, 88
Sukotyro	— I. 175

T

Tapiir	— I. 163
Tiger	— I. 277
——— Hunting	— I. 284

U

| Unicorn | — I. 155 |
| Urchin | — II. 234 |

V

	Vol. Page
Vampire	— II. 307
Vansire	— II. 51
Vicunna	— I. 136
Vison	— II. 51

W

WALRUS	— II. 266
Warree	— I. 141
Water Elephant	— I. 157
WEESEL	— II. 33
Wolf	— I. 248
Wolverene	— II. 8

Y

| Ysarus | — I. 72 |

Z

Zebra	— I. 13
Zerda	— I. 267
Zibet	— II. 72
Zorrina	— II. 66

INDEX

OF THE
NAMES of QUADRUPEDS,
IN THE
ANCIENT CLASSIC WRITERS,
IN THE
WORKS of M. DE BUFFON,
AND IN OTHER AUTHORS.

A

Name	Vol.	Page
ABBADOS —	I. -	154
Acanthion	II.	122
Addax	I. -	89
Adil	I. -	261
Adimain	I. -	40
Adive	I. -	261
Æegagrus	I. -	57
Agouti	II.	94
Ahu	I. -	93
Aï	II.	240
Aigrette	I. -	207
Akouchy	II.	95
Alagh-daagha	II.	169
Alce	I. -	105
Algazel	I. -	77
Allo-camelus	I. -	133
Allouatte	I. -	215
Alpaco	I. -	137
Amboimenes	I. -	230
Anak el Ard	I. -	306
Andira-guacu	II.	308

Name	Vol.	Page
Ane	I. -	8
Anta	I. -	163
Antelope	I. -	68
Apar	II.	246
Aper	I. -	140
Aperea	II.	90
Aquiqui	I. -	214
Arabata	I. -	215
Arctomys	II.	131
Arctopithecus	II.	240
Argali	I. -	44
Arnæb	II.	98
Αρκτος	II.	1
Armadillo	II.	246
Afhnoko	II.	92
Afpalax	II.	216
Affapanick	II.	153
Atterfoak	II.	279
Aurochs	I. -	17
Axis	I. -	117
Aye, Aye	II.	142
Ayotochtli		

VOL. II.　　　　　　　　U u

INDEX OF CLASSICAL NAMES, &c.

	Vol. Page		Vol. Page
Ayotochtli	II. 248	Brebis	I. - 37
		Bubalus	I. 28, 102
B		Bubel	I. - 24
Babouin	I. - 194	Bucula	I. - 103
Baby-rouffa	I. - 148	Buffle	I. - 28
Bafwer	II. 114	Bufelaphus	I. - 102
Baieu	I. - 122		
Bar	II. 1	**C**	
Baraba	II. 213	Caaigora	I. - 147
Barbarefque	II. 150	Cabiai	II. 88
Barbaftelle	II. 319	Cabionora	II. 88
Barrys	I. - 180	Cachicame	II. 248
Becheti	I. - 132	Cagui	I. 222, 224
Behemoth	I. - 157	Caitaia	I. - 220
Bekker el Wafh	I. - 102	Callitriche	I. - 203
Belette	II. 33	Callitrichus	I. - 203
Belier	I. - 40	Camelus	I. - 132
Beluga	II. 301	Camelo pardalis	I. - 65
Bey	I. - 212	Campagnol	II. 205
Biber	II. 114	Cani-apro-lupo-vulpes	I. - 272
Bievre	II. 114	Capefch	I. - 265
Biggel	I. - 83	Capibara	II. 88
Biorn	II. 1	Caprea	I. 96, 120
Bifon d'Amerique	I. - 23	Capreolus	I. - 120
Biur	II. 114	Capricorne	I. - 55
Blaireau	II. 14	Caracal	I. - 305
Bobak	II. 131	Caraco Mus	II. 181
Bobr	II. 114	Carcajou	II. 8
Bocht	I. - 132	Cariacou	I. - 122
Boern-doefkie	II. 157	Caribou	I. - 111
Bœuf	I. - 16	Carigueya	II. 20
Bonnet Chinois	I. - 209	Cariguibeiu	II. 79
Bofchratte	II. 90	Caftor	II. 114
Βεϛαλϴ	I. - 102	Cataphractus	II. 246
Bouc	I. - 60	Cavia, cobaya	II. 89
d' Angora	I. - 61	Cavia, genus	II. 88
d' Afrique	I. - 63	Cay	I. - 219
d' Juda	I. - 63	Cayopollin	II. 24
Baumriitter	I. - 295	Cemas	I. - 85
Bouquetin	I. - 55	Cerf	I. - 114
Boury	I. - 21	Chacal	I. - 261
Βεϛ αγριος	I. - 28	Chameau	I. - 132
Bradypus	II. 240	Chamois	I. - 44
Brandhirtz	I. - 116	Chat	I. - 295

Chat

INDEX OF CLASSICAL NAMES, &c.

	Vol.	Page
Chat d' Angora	I. -	296
d' Efpagne	I. -	296
Chat-pard	I. -	300
Chaus *Plinii*	I. -	301
Chauve-fouris	II.	311
Cheropotamus	I. -	156
Cherofo	II.	186
Cheval	I. -	1
Chevre	I. -	60
Chevreuil	I. -	120
Chevrotain d'Afrique	I. -	81
de Guinèe	I. -	82
des Indies	I. -	127
Chien	I. -	235
Chimpanzee	I. -	180
Chinchè	II.	65
Chinchimen	II.	82
Chinchilla	II.	196
Chomik	II.	206
Choras	I. -	188
Cirquinçon	II.	250
Citillus	II.	135
Civette	II.	70
Clap–Myfs	II.	279
Coaita	I. -	216
Coafe	II.	62
Coati	II.	12, 61
Cochon	I. -	140
Cochon d'Inde	II.	89
Coendu	II.	125
Coymatl	I. -	147
Colus	I. -	98
Condoma	I. -	88
Conepate	II.	64
Coquallin	II.	147
Corine	I. -	101
Coudous	I. -	78
Couguar	I. -	289, 290
Coyotl	I. -	257
Coypu	II.	177
Crabier	II.	24
Cricetus	II.	206
Crocuta	I. -	272
Cuandu	II.	124

	Vol.	Page
Cuetlatchtli	I. -	250
Cuguaca-apara	I. -	122
Cugacuara	I. -	289
Cuguacu-ete	I. -	126
Cugacuarana	I. -	289
Culpeu	I. -	258
Cuniculus	II.	103
Cynocephalus	I. -	186

D
Dachs	II.	14
Dain	I. -	113
Dama, Antelope	I. -	85
Daman Ifrael	II.	92
Dam-tanhirfch	I. -	113
Dandoelana	II.	140
Daniel	I. -	113
Dant	I. -	21
Dæfman *	II.	221
Dafypus	II.	246
Djammel	I. -	129
Diane	I. -	201
Dof, Dof-hiort	I. -	113
Dorcas	I. 92, 102, 120	
Douc	I. -	211
Dromedaire	I. -	129
Dferen	I. -	96
Dfhikketaei	I. -	4
Dubha	I. -	270
Dugon	II.	269
Dzik	I. -	140

E
Echinus Terreftris	II.	234
Ecureuil	II.	138
Eichhorn	II.	138
Elan	I. -	105
Elephant, Elephas	I. -	165
Elk	I. -	105
Encourbert	II.	247
Engalla	I. -	144
Erdzeifl	II.	205
Erinaceus	II.	234, 248
Exquima	I. -	201

* *De Buffon,* x. 1. *tab.* 11.

Fara

INDEX OF CLASSICAL NAMES, &c.

	Vol.	Page		Vol.	Page
F			Guerlinguets	II.	162
Fara	II.	18	Guevei	I. -	82
Felis Catus	I. -	295	Guib	I. -	81
Feuille	II.	309	Guillino	II.	120
Fial racka	I. -	255	Gulo	II.	10
Fiber	II.	114	Gundi	II.	137
Filander	II.	22			
Filfrefs	II.	10	**H**		
Fifkatta	II.	65	Hamfter	II.	206
Fifhtal	I. -	53	Handl	I. -	150
Flader-mus	II.	319	Hardlooper	I. -	144
Fong kyo fo	I. -	133	Hafe	II.	98
Foffane	II.	75	Hærbe	II.	234
Fouine	II.	41	Heriffon	II.	234
Fourmillier	II.	260	Hermine	II.	35
Foyna	II.	41	Hippelaphus	I. -	115
Fret	II.	40	Hippopotamus	I. -	157
Fuchs	I. -	251	Hirax	II.	92
Furo	II.	40	Hirco cervus	I. -	52
			Hiort	I. -	114
G			Hirfch	I. -	114
Ganfud	II.	234	Hoang-yang	I.	96
Galera	II.	53	Huanucu-Llama	I. -	133
Galgopithecus	I. -	234	Hugiun	I. -	129
Gazelle	I. 75, 77, 89		Hyæna	I. -	270
Gemfe	I. -	72	Hydrochærus	II.	88
Genette	II.	74	Hyftrix	II.	122
Gerbo	II.	170			
Ghainoûk	I. -	27	**I**		
Gibbon	I. -	184	Jærven	II.	10
Giraffe	I. -	65	Jaguar	I. 284, 286	
Glis	II.	158	Jaguarete	I. -	290
Glutton	II.	10	Jagura	I. -	286
Glouton	II.	8	Jarf	II.	10
Gnou	I. -	70	Javaris	I. -	147
Gornoftay	II.	35	Ibex	I. -	55
Grafkin	II.	138	Ichneumon	II.	54
Grimme	I. -	81	Jelen	I. -	114
Grifon	II.	52	Jez	II.	234
Guachi	II.	80	Jerboa	II.	164
Guanaco	I. -	138	Igel, Igelkott	II.	234
Guanque	II.	183	Ignavus	II.	240
Guareba	I. -	214	Indri	I. -	228
Guepard *	I. -	284	Jocko	I. -	180

* *De Buffon*, xiii, 254. The fame with the Hunting Leopard, *No.* 184.

Iππος

INDEX OF CLASSICAL NAMES, &c.

	Vol. Page
Ιππος ποταμιος	I. - 157
Irabubos	II. 88
Isatis	I. - 255

K

	Vol. Page
Kabarga	I. - 124
Kabassou	II. 249
Kalan	II. 83
Kangaru	II. 29
Kanin	II. 103
Karagan	I. - 252
Kassigiak	II. 270
Kattlo	I. - 301
Κηϐ⊙.	I. - 210
Kenlie	I. - 265
Kevel	I. - 92
Kidang	I. - 119
Kinkajou	II. 60
Kob	I. - 104
Koba	I. - 103
Kolonnok	II. 39
Kot draki	I. - 295
Koulan	I. - 8
Kret	II. 229
Kron-hiort	I. - 114
Krietsch	II. 206
Krylatca	II. 279
Kuna	II. 41

L

	Vol. Page
Llama	I. - 133
Lacertus	II. 252
Laderlap	II. 319
Lame	II. 285
Lant	I. - 21
Lapin	II. 103
d' Angora	II. 104
Lar	I. - 185
Latax	II. 80
Lemur	I. - 227
Lemmar, Leming	II. 198
Lemni	II. 214
Leo	I. - 274
Leopard	I. - 282
Lepus	II. 98

	Vol. Page
Lerot	II. 159
Lerwee	I. - 53
Leucoryx	I. - 76
Levrier	I. - 241
Lidmeè	I. - 91
Lievre	II. 98
Lion	I. - 274
Loir	II. 158
Loris	I. - 228
Loup	I. - 248
de Mexique	I. - 250
Loup-Cervier	I. - 301
Loup-Renard	I. - 257
Loutre	II. 77
Lowe	I. - 274
Lupus	I. - 248
Lummick	II. 198
Lutra	II. 77
Lux	I. - 301
λυγξ	I. - 301
Lynx	I. - 301

M

	Vol. Page
Maucauco	I. - 227
Machlis	I. - 105
Mafutiliqui	II. 66
Magot	I. - 186
Magu	I. - 213
Maimon	I. - 190
Malbrouck *	I. - 201
Mammouth	I. - 172
Manati	II. 292
Mandril	I. - 190
Mangabey	I. - 204
Mangouste	II. 54
Manicou	II. 18
Manipouris	I. - 163
Manis	II. 252
Manul	I. - 294
Mapach	II. 12
Maraguao	I. - 292
Mard	II. 41
Margay	I. - 292
Marikina	I. - 225
Mariputa	II. 66

* De Buffon, xiv. 224. tab. xxix. A variety of our *Egret*, No. 119.

Marmose

INDEX OF CLASSICAL NAMES, &c.

	Vol.	Page		Vol.	Page
Marmofe	II.	23	Nanguer	I. -	85
Marmotte	II.	128	Neitfek	II.	278
Martes, Marte	II.	41, 42	Nems	II.	54
Mejangan Banjoe	I. -	118	Niedzwiedz	II.	1
Meles	II.	14	Nietferfoak	II.	279
Meminna	I. -	127	Nil-ghau	I. -	83
Mico	I. -	226	Noctule	II.	317
Mocawk	I. -	230	Noerza	II.	80
Mococo	I. -	230	Norka	II.	80
Molle	II.	205	Νυκτερις	II.	319
Monax	II.	130			
Mone	I. -	210	**O**		
Monèa	I. -	207	Ocelot	I. -	287
Mongooz	I. -	229	Ochs	I. -	16
Moofe	I. -	107	Odobenus	II.	266
Morfe	II.	266	Ogotona	II.	109
Morfkuia Korawa	II.	292	Onager	I. -	11
Mouffettes *	II.	62	Once	I.	285, 290
Mouflon	I. -	44	Ondatra	II.	119
Mouftac	I. -	205	Opeagha	I. -	14
Mouton de Barbarie	I. -	41	Ophion	I. -	44
Mufro	I. -	46	Orang Outang	I. -	180
Mulet	I. -	8	Oreillar	II.	320
Μυγαλη	II.	224	Oreotragus	I. -	79
Mullvad	II.	229	Orignal. See Elk.		
Mulot	II.	184	Oryx	I. -	76
Munt-jak	I. -	119	Oftrowidz	I. -	301
Murmelthier	II.	128	Ouaikarè	II.	240
Mus Alpinus	II.	128	Ouanderou	I. -	198
Mus	II.	184	Ouarine	I. -	214
Mufaraigne	II.	224	Ouiftiti	I. -	224
Mus Araneus	II.	224	Ourebi	I. -	79
Mufc	I. -	124	Ourico	II.	124
Mufcardin	II.	160	Ours	II.	1
Mufimon	I. -	44	Ours blanc de mer	II.	5
Mufquafh	II.	119	Ours marin	II.	281
Muffafcus	II.	119	Ovis	I. -	37
Muftela	II.	33			
Myrmecophaga	II.	256	**P**		
			Paca	II.	91
N			Pacaffe	I. -	78
Nabbmus	II.	224	Paco, Pacos	I. -	136
Nagor	I. -	86	Palatine	I. -	200

* M. De Buffon's generic name for the Polecats which exhale fo peftilential a vapour.

Palmifte

INDEX OF CLASSICAL NAMES, &c.

	Vol.	Page
Palmiste	II.	149
Pangolin	II.	253
Panthera, Panthere	I. -	280
Papio	I. -	188
Παρδαλις	I.	280, 285
Pardus	I. -	280
Paresseux	II.	240
Pasan, Pasan	I.	57. 75
Patas	I. -	208
Pecary	I. -	147
Pekan	II.	51
Pelander Aroe	II.	21
Perchal	II.	179
Pere	I. -	12
Perugusna	II.	38
Petit Gris	II.	144
Phalanger	II.	27
Phatagin	II.	252
Philandre	II.	27
Philodotus	II.	252
Phoca	II.	270
Phoque	II.	270
Pichou	I. -	292
Piloris	II.	97
Pilosello	II.	74
Pinche	I. -	225
Pipistrelle	II.	318
Pissay	I. -	127
Πιθηκ⊙, Pitheque	I. -	183
Platogna	I. -	113
Platyceros	I. -	113
Poephagus	I. -	27
Poulatouche	II.	155
Pongo	I. -	180
Porc-epic	II.	122
Προξ	I. -	113
Przewiaska	II.	38
Pteropus	II.	304, 308
Puma	I. -	289
Putois	II.	37
Putorius	II.	37

Q

	Vol.	Page
Quagga	I. -	14

	Vol.	Page
Quahtechalotl-thlitic	II.	145
Quapizotl	I. -	147
Quato	I. -	216
Quauhtla	I. -	147
Quil, Quirpele	II.	54
Quojas Morrou	I. -	180
Quoll	II.	69
Quouata	I. -	216
Quumbengo	I. -	272

R

	Vol.	Page
Radjur	I. -	120
Raef	I. -	251
Rangier	I. -	111
Rangwo	II.	105
Rat	II.	176
d'Eau	II.	182
de Madagascar	I. -	233
Ratel	II.	66
Raton	II.	12
Rattepenade	II.	320
Rein Deer	I.	111
Renard	I. -	251
Renne	I. -	111
Rennthier	I. -	111
Rhen	I. -	111
Rhinoceros	I. -	150
Rillow	I. -	209
River Paard	I. -	157
Roloway	I. -	200
Root	I. -	213
Roselet	II.	35
Rosmarus	II.	266
Rosomak	II.	10
Rougette	II.	304
Rouffette	II.	304
Rukkaia	II.	140
Rupicapra	I.	44, 72
Rusla	II.	112
Rys	I. -	308

S

	Vol.	Page
Saca	I. -	294
Saccawinkee	I. -	222
Sagouin		

INDEX OF CLASSICAL NAMES, &c

	Vol.	Page
Sagouin	I. -	224
Sai	I. -	218
Saiga	I. -	98
Saimiri	I. -	220
Sajou	I. -	217
Saki	I. -	222
Sanglier	I. -	140
de Capvert	I. -	144
Sanglin	I. -	224
Sapajou	I. -	222
Saragoy	II.	20
Saricovienne	II.	82
Sarigue	II.	20
Sarlyk	I. -	27
Σαθεριον	II.	45
Satyrus	I. -	180
Scenoontung	I. -	122
Schakal	I. -	261
Schwein	I. -	140
Sciurus	II.	138
Semlanoi Saetſhik	II.	112
Serotine	II.	317
Serval	I. -	301
Shitnik	II.	190
Sial	II.	270
Siegen Bock	I. -	60
Sifac	I. -	211
Simia	I. -	183
Siwutſcha	II.	288
Siya	II.	79
Siyah Ghuſh	I. -	305
Skrzeczek	II.	206
Slepez	II.	214
Sno-mus	II.	33
Sobol	II.	47
Sogur	II.	131
Songar	II.	212
Sorex	II.	224
Souris	II.	184
Souſlik	II.	135
Speck-maus	II.	319
Spring-bock	I. -	94
Springen Haas	II.	170

	Vol.	Page
Squilachi	I. -	261
Squinaton	I. -	122
Steinbock	I. -	55
Stink bingſem	II.	66
Stock	I. -	150
Strepſiceros	I. 38, 88	
Suhak	I. -	98
Suiſſe, Ecureuil	II.	157
Sumxi	I. -	296
Surikate	II.	57
Surk	II.	229
Surmulot	II.	178
Sus	I. -	140
Sus Aquaticus	I. -	163
Swiftch	II.	131

T

	Vol.	Page
Taguan	II.	151
Tajacu	I. -	147
Tajibi	II.	18
Taiſon	II.	14
Talapoin	I. -	206
Talpa	II.	229
Tamandua	II.	256
Tamanoir	II.	256
Tamarin	I. -	223
Tanrec	II.	236
Tapeti	II.	107
Tapir	I. -	163
Tapoa Tafa	II.	69
Tarandus	I. -	111
Tardigradus	II.	240
Tarſier	I. -	231
Tartarin	I. -	194
Tatou	II.	248
Tatu apara	II.	246
Tatuete	II.	248
Taupe	II.	229
Taupe dorèe	II.	231
Taureau	I. -	16
Taxus	I. -	270
Tayra	II.	53
Tchorz	II.	37
Tegoulichitck	II.	194

Tegul

INDEX OF CLASSICAL NAMES, &c.

	Vol.	Page
Tegul		
Temamaçama	I. -	103
Tendrac	II.	236
Tenlie	I. -	265
Tepe Maxlaton	I. -	292
Teutlalmaçama	I. -	122
Tgao	I. -	157
Thous	I. -	267
Tigris, Tigre	I. -	277
Tlaloceloti	I. -	287
Tla-coozelotl	I. -	287
Tlalmototli	II.	149
Tlaquatzin	II. 18.	124
Tolai	II.	104
Touan	II.	34
Tragelaphus	I. -	52
Trago-Camelus	I. -	83
Tragulus	I. -	124
Tretretretre	I. -	191
Trichechus Rofmarus	II.	266
——— Manatus	II.	297
Tfchotfchot	II.	
Tfitsjan	II.	135
Tucan	II.	223
Tzeiran	I. -	74

V

Vache Marine	II.	266
de Tartarie	I. -	24
Ϋαλϗα	I. -	270
Vampire	II.	308
Vanfire	II.	51
Vary	I. -	229
Varia	I. -	280
Vavi	I. -	261
Veldratte	II.	90
Verdadeiro	II.	248
Vefpertilio	II.	304
Vigogne, Vicunna	I. -	136
Vifon	II.	51
Viverra	II.	40
Viverra tigrina	I. -	298

	Vol.	Page
Ulf	I. -	248
Unau	II.	242
Uncia	I. -	282
Vormela	II.	210
Urigne	II.	290
Urfon	II.	126
Urfus	II.	1
Urus	I. -	16
Ϋϛ τετραϗερος	I. -	148
Ϋϛριξ	II.	122
Vulpes	I. -	251
Utfuk	II.	277
Utter	II.	77

W

Walrus	II.	266
Warg	I. -	248
Warglo	I. -	301
Warree	I. -	141
Weefel	II.	33
Wettfk	I. -	150
Whang, Yang	I. -	96
Wha Tapoua Row	II.	13
Wiewiorka	II.	138
Wirrebocarra	I. -	126
Whydra	II.	77

X

Χοιροπιθηϗος	I. -	187
Xoloitzcuintli	I. -	250

Y

Yerboa	II.	29
Yltis	II.	37
Yfard, Yfarus	I. -	72
Yzquiepatl	II.	62

Z

Zbik	I. -	295
Zebre, Zebra	I. -	13
Zebu	I. -	21
Zecora		

INDEX OF CLASSICAL NAMES, &c.

		Vol. Page			Vol. Page
Zecora	———	I. - 13	Zibet	———	II. 72
Zenik	———	II. 193	Zizel	———	II. 135
Zerda	———	I. - 267	Zobela	———	II. 43
Zibeline.	———	II. 43	Zorrina *	———	II. 66

* *De Buffon,* xiii. 302. *tab.* xli.

THE END.

ERRATUM.

Nos. 273 and 276, Pages 70 and 72, being the fame animal, the reader is defired to correct this miftake.